Advanced Textbooks in Control and Signal Processing

Springer

London
Berlin
Heidelberg
New York
Hong Kong
Milan
Paris
Tokyo

Series Editors
Professor Michael J. Grimble, Professor of Industrial Systems and Director
Professor Michael A. Johnson, Professor of Control Systems and Deputy Director
Industrial Control Centre, Department of Electronic and Electrical Engineering,
University of Strathclyde, Graham Hills Building, 50 George Street, Glasgow G1 1QE, U.K.

K.M. Hangos, J. Bokor and G. Szederkényi

Analysis and Control of Nonlinear Process Systems

With 42 Figures

 Springer

K.M. Hangos, PhD, DSci
J. Bokor, PhD, DSci
G. Szederkényi, PhD
Systems and Control Laboratory, Computer and Automation Institute,
Hungarian Academy of Sciences, H-1516 Budapest, PO Box 63, Kende u. 13-17,
Hungary

British Library Cataloguing in Publication Data
A catalogue record for this book is available from the British Library

Library of Congress Cataloging-in-Publication Data
Hangos, K. M. (Katalin M.)
    Analysis and control of nonlinear process systems / K.M. Hangos, J. Bokor [sic], and G. Szederkényi.
        p. cm. - - (Advanced textbooks in control and signal processing, ISSN 1439-2232)
    ISBN 1-85233-600-5 (alk. Paper)
        1. Process control. 2. Nonlinear control theory. I. Bokor, J. József), 1948- II.
    Szederkényi, G. (Gàbor), 1975- III. Title. IV. Series.
    TS156.8.H34 2004
    629.8'36-dc22
                                    2003065306

ISSN 1439-2232
ISBN 1-85233-600-5 Springer-Verlag London Berlin Heidelberg
a member of BertelsmannSpringer Science+Business Media GmbH
springeronline.com

Typesetting: Electronic text files prepared by authors
Printed and bound in the United States of America
69/3830-543210 Printed on acid-free paper SPIN 10868248

For God had not given us the spirit of fear; but of power, and of love, and of a sound mind.

II. Timothy 1.6

Foreword

Process systems constitute a key aspect of human activity that is closely linked to the three pillars of sustainable development: Economic competitiveness, Social importance (employment, quality of life) and Environmental impact. The future economic strength of societies will depend on the ability of production industries to produce goods and services by combining competitiveness with quality of life and environmental awareness. In the combined effort to minimize waste through process integration and to optimally operate the constructed processes nonlinear behaviours are being exploited. Thus there will be an increasing need for nonlinear process theory to systematically deal with the relatively complex nonlinear issues that appear with the increasing process systems complexity dictated by our technological capability and the competitive demands.

This book serves as a most promising source that combines process systems engineering with nonlinear systems and control theory. This combination is carried through in the book by providing the reader with references to linear time invariant control theory. The nonlinear passivity theory constitutes a particularly promising contribution that is illustrated on problems of relatively low dimensionality.

The successful establishment of the state-of-art in nonlinear process systems control in a concise textbook represents a laudable contribution to process systems theory for the benefit of future graduate students and researchers and hopefully also for the benefit of human activity.

Lyngby, July 2003

Professor Sten Bay Jørgensen
Director of CAPEC (Computer Aided Process Engineering Center)
Department of Chemical Engineering
Technical University of Denmark
Lyngby, Denmark

Series Editors' Foreword

The topics of control engineering and signal processing continue to flourish and develop. In common with general scientific investigation, new ideas, concepts and interpretations emerge quite spontaneously and these are then discussed, used, discarded or subsumed into the prevailing subject paradigm. Sometimes these innovative concepts coalesce into a new sub-discipline within the broad subject tapestry of control and signal processing. This preliminary battle between old and new usually takes place at conferences, through the Internet and in the journals of the discipline. After a little more maturity has been acquired by the new concepts then archival publication as a scientific or engineering monograph may occur.

A new concept in control and signal processing is known to have arrived when sufficient material has evolved for the topic to be taught as a specialised tutorial workshop or as a course to undergraduate, graduate or industrial engineers. *Advanced Textbooks in Control and Signal Processing* are designed as a vehicle for the systematic presentation of course material for both popular and innovative topics in the discipline. It is hoped that prospective authors will welcome the opportunity to publish a structured and systematic presentation of some of the newer emerging control and signal processing technologies in the textbook series.

As most of the problems from linear control analysis have found solutions, advances in future control performance will come from accommodating the non-linear nature of many processes more directly. This is a challenge facing many areas of control engineering. In the process industries there is a fair amount of non-linear model information and the task is to find ways to exploit this knowledge base. On the other hand the analysis of non-linear systems *per se* is reasonably well developed but in many cases the move to more routine application of these techniques still remains to be taken. We believe it is only by having the utility and advantages of non-linear control demonstrated in practical applications that the non-linear control paradigm will begin to make a contribution to control engineering.

Process control is one area where there is the possibility of demonstrating a major advance through the use of non-linear control. Tackling this challenge we are pleased to have this textbook by Katalin Hangos, József Bokor, and Gábor Szederkényi on "Analysis and control of non-linear process systems" in the *Advanced Textbooks in Control and Signal Processing* series. It is a text based on past course experience and care has been taken to enhance the accessibility of the material with nice pedagogical features like special indexes, boxed important definitions and end-of-chapter exercises. The underlying rigour of the non-linear analysis has however been preserved. The book is suitable for graduate and postgraduate courses in process systems engineering and for self-study at that level. It is our hope that this textbook will contribute to the more widespread acceptance of non-linear control in applications.

M.J. Grimble and M.A. Johnson
Industrial Control Centre
Glasgow, Scotland, U.K.
Summer 2003

Contents

List of Definitions

List of Examples

Acknowledgements

The authors are indebted to many of their colleagues for encouragements, discussions and constructive criticism. Tibor Vámos (Computer and Automation Research Institute of the Hungarian Academy of Sciences, Hungary), Sten Bay Jørgensen (CAPEC Centre, Technical University of Denmark, Denmark), Ian Cameron (CAPE Centre, The University of Queensland, Australia) and Ferenc Szigeti (Venezuela) have influenced our ideas the most.

We are also grateful to our students and collaborators with whom we worked together on projects related to the subject of the present book: to Piroska Ailer, Attila Magyar, Huba Németh, Tamás Péni and Barna Pongrácz, to mention just a few.

The related research has been sponsored by the Hungarian National Research Fund through grants $T032479$ and $T042710$, which is gratefully acknowledged. The third author acknowledges the support of the Bolyai János Scholarship of the Hungarian Academy of Sciences.

1. Introduction

Process control is a traditional area in process systems engineering which is of great practical importance. Control usually involves other related fields, like dynamic modeling, identification, diagnosis, *etc.*

The majority of practical control applications are mainly based on PID loops; there are only a few advanced control systems, either for operating units (equipment) with complex or unstable dynamic behavior, or plantwide optimizing control.

> *It is widely known in process systems engineering that almost all process systems are nonlinear in nature.*

Therefore advanced process control should necessarily use nonlinear control techniques.

1.1 A Brief Overview of Nonlinear Process Control

Elementary or introductory control courses for both control (electrical) and process engineers are almost entirely based on *linear systems*; this is what we all start with. The reason for this is twofold. First of all, there are relatively simple closed analytical solutions to many control problems (including LQR and pole-placement controller design, Kalman-filtering, model parameter and structure estimation, *etc.*), so the linear theory is nice, transparent and feasible. On the other hand, practical applications are also based on linear or linearized models in most cases and handle nonlinearities only when it is absolutely unavoidable.

The common way of controlling process systems with strong nonlinear character is to apply model-based predictive controllers where a detailed dynamic process model is used in an optimization framework. The popularity of model-based predictive control is partially explained by the fact that it fits so well into the "culture" of process systems engineering: it uses traditional dynamic process models which are usually available for design and/or simulation purposes. At the same time, model-based predictive control is being criticized by control engineers because of its lack or weakness of theoretical background, having no guarantee of convergence, stability, robustness, *etc.* in the general case.

Modern heuristic black-box-type control approaches, such as neural nets and fuzzy controllers, have also appeared recently even in industrial practice. At the same time, the results and approaches of modern nonlinear control theory have not earned acceptance in the field of process control. There are two reasons. Firstly, these techniques require an advanced mathematical background and skills, which are rarely taught to process engineers. Secondly, modern nonlinear control methods are computationally hard, and are only feasible for small-scale systems in the general case.

These problems with nonlinear control techniques applied in the general case indicate that a solid knowledge of the special characteristics of the nonlinear system in question may significantly help in developing nonlinear controllers for process systems with reasonably realistic complexity. Therefore, any work in the area of nonlinear process control should be based on an interdisciplinary approach that integrates the results and techniques of process systems engineering with nonlinear systems and control theory. The interdisciplinary nature of this approach behooves us to present an overview of the existing literature in both fields from a special integrating viewpoint.

There are excellent and widely used textbooks where the *nonlinear analysis and control techniques* are presented, such as:

1. H. Nijmeijer and A.J. Van der Schaft (1990) *Nonlinear Dynamical Control Systems*, Springer.
2. A. Isidori (1995) *Nonlinear Control Systems I.-II. (Communication and Control Engineering Series)*, Springer.
3. A.J. van der Schaft (1999) *L2-Gain and Passivity Techniques in Nonlinear Control (Communication and Control Engineering Series)*, Springer.

The above textbooks intend to cover a range of nonlinear analysis and control techniques in an abstract mathematical way, concentrating on the aspects relevant for refining or developing further the techniques. Also, the intended readers are graduate and postgraduate students specializing in control engineering or in applied mathematics. Therefore these books do not meet the needs of process engineering students interested in the application of these methods for process systems.

There are also application oriented textbooks available on the market dealing with nonlinear control, such as [10], [66] and [60]. These books cover a wide range of application areas, therefore they cannot take the specialities of process systems into account and do not build on the engineering knowledge readily available in the field.

1.2 Aims and Objectives

This textbook has been written for graduate and postgraduate students with a process engineering background. It aims to bridge the gap between process systems engineering and advanced nonlinear control theory by:

- providing the necessary mathematical preliminaries for graduate and postgraduate students with a process engineering background,
- presenting the relevant and promising theory and methods for analyzing and controlling nonlinear process systems,
- emphasizing the importance and use of process knowledge, mainly process models developed from first engineering principles, for obtaining feasible and effective special cases of the general methods.

The textbook deals with the basic concepts and with the most promising tools and techniques in nonlinear process analysis and control illustrated by simple examples and tutorial material.

The notions and techniques are always introduced and illustrated in the standard finite dimensional linear time-invariant continuous case, which serves as a basis for an extension to the nonlinear case. This way, the necessary links are also established with more widely known material, which makes the understanding of the concepts and methods a lot easier.

The Level of the Text. The level corresponds to graduate or postgraduate courses in process systems engineering.

The interdisciplinary and rapidly developing nature of the topic as well as the broad and diverse background of the potential readers requires us to restrict the prerequisite knowledge to a necessary minimum. Only basic higher mathematics common in engineering courses, such as linear algebra and elementary calculus are assumed. A solid knowledge of process modeling and control are advisable.

The advanced mathematical tools and notions we build upon are summarized in **Appendix A** of the textbook.

Learning Aids. This book is primarily a textbook. Therefore we provide special learning tools in order to make its use more comfortable for both lecturers and students of a course in analysis and control of nonlinear process systems. These include:

- an **Index** containing the special terms, definitions, methods and other important knowledge elements in the book,
- a **List of Definitions**, which is a special index for the most important terms and notions in the book,
- simple worked in-text examples, which have the following format:

Example 1.2.1 (Example of examples)
A simple example

....

- a **List of Examples** of the above which is another special index,
- learning aid sections, such as **Summary, Questions and Application Exercises** at the end of each chapter,
- an additional learning aid section **Further Reading** at the end of each non-introductory chapter (Chapters 5–12),
- a simple typographical scheme to distinguish important terms, definitions and statements in the text in the form:

> *This is an important knowledge item.*

1.3 The Road Map of the Book

The material is divided into three logical parts:

- *Control-oriented modeling of concentrated parameter nonlinear systems* (Chapters 1-5).
  The basic notions of systems and signals are presented first in Chapter 2. Thereafter linear and nonlinear state-space models are discussed (Chapter 3). The construction and the special structure of dynamic process models is described in Chapter 4. Chapter 5 is devoted to input–output models important for some of the analysis and control techniques later on,
- *Nonlinear system analysis methods and tools* (Chapters 6–8).
  The most important nonlinear analysis methods for controllability, observability and stability are described first (Chapters 6 and 7). Chapter 8 is devoted to passivity analysis and the Hamiltonian description, which are powerful concepts with an important physical basis in the case of process systems,
- *Nonlinear feedback control* (Chapters 9–12).
  An introductory chapter (Chapter 9) deals with the basic notions and techniques for state feedback control, in particular pole-placement and LQR control of linear time-invariant systems. Thereafter separate chapters deal with the most important special techniques for nonlinear process control: feedback and input–output linearization, passivation and stabilization and loop-shaping based on the Hamiltonian view.

Finally, the necessary mathematical preliminaries not contained in a standard process engineering curriculum, namely coordinate transformations, norms, Lie-derivatives and -products, distributions as well as co-distributions are summarized in **Appendix A**.

The road map of the book, which shows using arrows the dependence of the material presented in the chapters, is depicted in Figure 1.1. Directed paths show how to proceed when taking a tour through the material: either by giving/taking a course or by self-learning.

The dashed arrows connecting Chapter 4 "Dynamic process models" to the other chapters indicate that each non-introductory chapter contains at least one section where process system case studies are described.

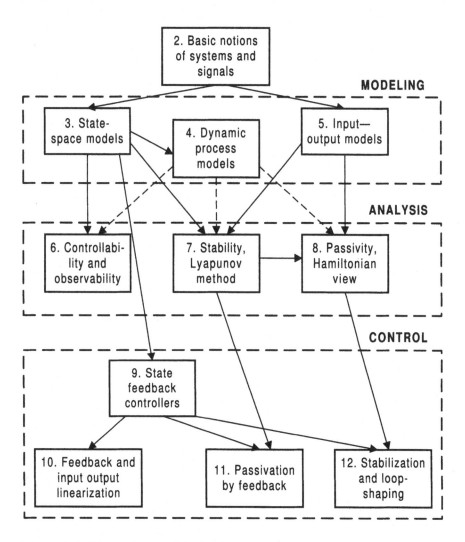

Figure 1.1. The road map of the book

The first draft version of this material was used for an intensive four-week postgraduate course taught for process engineers in the CAPEC Center, Department of Chemical Engineering, Technical University of Denmark (Lyngby, Denmark) in 2000.

Part of the material has been used for an elective course entitled "Modern control methods" taught for information engineers at the University of Veszprém, Hungary since 2003.

2. Basic Notions of Systems and Signals

Signals and systems are basic notions of systems and control theory. Therefore, we briefly summarize the most important concepts of the mathematical description of signals and systems in this introductory chapter in the form they are used later on.

The material in this chapter is presented in two sections:

- *Signals*
  Not only the definition and classification of signals, operations on signals and the notion of signal spaces are given but the most important special signal types are also introduced.
- *Systems*
  The most important system classes are defined based on the abstract notion of systems together with the characterization of their input–output stability.

2.1 Signals

Signals are the basic elements of mathematical systems theory, because the notion of a system depends upon them.

2.1.1 What is a Signal?

Generally, a *signal* is defined as any physical quantity that varies with time, space or any other independent variable(s). A signal can be a function of one or more independent variables.

A longer and more application-oriented definition of signals taken from [50] is the following: "Signals are used to communicate between humans and between humans and machines; they are used to probe our environment to uncover details of structure and state not easily observable; and they are used to control energy and information."

Example 2.1.1 (Simple signals)

The following examples describe different simple signals.

- Let us suppose that the temperature x of a vessel in a process plant is changing with time (measured in seconds) in the following way:

$$x : \mathbb{R}_0^+ \mapsto \mathbb{R}, \quad x(t) = e^{-t}$$

We can see that t is the independent variable (time) and x is the dependent variable (see Figure 2.1 (/a)).

- Let us assume that we can observe the temperature x in the first example only at integer time instants (seconds). Let us denote the observed temperature by y, which therefore can be defined as:

$$y : \mathbb{N}_0^+ \mapsto \mathbb{R}, \quad y[n] = e^{-n}$$

(see Figure 2.1 (/b)).

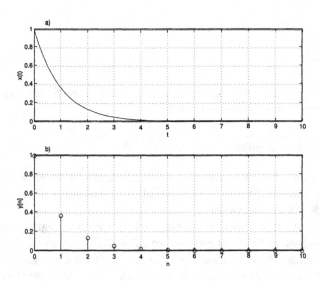

Figure 2.1. (/a) Simple continuous time signal and (/b) its discrete time counterpart

- A complex-valued signal with a complex independent variable is, *e.g.*

$$X : \mathbb{C} \mapsto \mathbb{C}, \quad X(s) = \frac{1}{s+1}$$

  which is actually the Laplace transform of the first signal x (see Subsection 2.1.4).
- Let us denote the temperature of a point (x, y, z coordinates in a Cartesian coordinate system) inside a room at a certain time instant t by $T(x, y, z, t)$. It's easy to see that T is a real–valued function of four independent variables and hence it maps from $\mathbb{R}^4$ to $\mathbb{R}$.
- An example of a vector-valued signal with three independent variables is the picture of a color TV where the intensity functions of the red, green and blue colors (I_r, I_g and I_b) form the vector

$$I(x, y, t) = \begin{bmatrix} I_r(x, y, t) \\ I_g(x, y, t) \\ I_b(x, y, t) \end{bmatrix}$$

  i.e. $I : \mathbb{R}^3 \mapsto \mathbb{R}^3$. The independent variables in this case are the screen coordinates (x and y) and time (t).

2.1.2 Classification of Signals

Signals are classified according to the properties of their independent and dependent variables.

Dimensionality of the Independent Variable. The independent variable can have one or more dimensions. The most common one-dimensional case is when the independent variable is time.

Dimension of the Dependent Variable (Signal). The signal value evaluated at a certain point in its domain can also be one or more dimensional.

Real–valued and Complex–valued Signals. The value of a signal can be either real or complex. The following question naturally arises: why do we deal with complex signals? The answer is that the magnitude and angle of a complex signal often has clear engineering meaning, which is sometimes analytically simpler to deal with.

Continuous Time and Discrete Time Signals. In systems and control theory, one usually has a one-dimensional independent variable set which is called *time* and denoted by T in the general case.

Continuous time signals take real or complex (vector) values as a function of an independent variable that ranges over the real numbers, therefore a continuous time signal is a mapping from a subset of $\mathbb{R}$ to $\mathbb{C}^n$, *i.e.* $T \subseteq \mathbb{R}$.

Discrete time signals take real or complex (vector) values as a function of an independent variable that ranges over the integers, hence a discrete time signal makes a mapping from $\mathbb{N}$ to $\mathbb{C}^n$, *i.e.* $T \subseteq \mathbb{N}$.

Bounded and Unbounded Signals. A signal $x : \mathbb{R} \mapsto \mathbb{C}$ is bounded if $|x(t)|$ is finite for all t. A signal that does not have this property is unbounded.

Periodic and Aperiodic Signals. A time-dependent real-valued signal $x : \mathbb{R} \mapsto \mathbb{R}$ is periodic with period T if $x(t+T) = x(t)$ for all t. A signal that does not have this property is aperiodic.

Even and Odd Signals. Even signals x_e and odd signals x_o are defined as

$$x_e(t) = x_e(-t)$$
$$x_o(t) = -x_o(-t)$$

Any signal is a sum of unique odd and even signals. Using $x(t) = x_e(t) + x_o(t)$ and $x(-t) = x_e(t) - x_o(t)$ yields $x_e(t) = \frac{1}{2}(x(t) + x(-t))$ and $x_o(t) = \frac{1}{2}(x(t) - x(-t))$.

2.1.3 Signals of Special Importance

Some signals are of theoretical and/or practical importance because they are used as special test signals in dynamic systems analysis.

Definition 2.1.1 (Dirac-δ or unit impulse function)
The Dirac-δ or unit impulse function is not a function in the ordinary sense. The simplest way it can be defined is by the integral relation

$$\int_{-\infty}^{\infty} f(t)\delta(t)dt = f(0) \tag{2.1}$$

where $f : \mathbb{R}_0^+ \mapsto \mathbb{R}$ is an arbitrary smooth function.

The unit impulse is not defined in terms of its values, but by how it acts inside an integral when multiplied by a smooth function f. To see that the area of the unit impulse function is 1, we can choose $f(t) = 1$ in the definition.

The unit impulse function is widely used in science and engineering. The following statements illustrate its role from an engineering point of view:

- impulse of current in time delivers a unit charge instantaneously to an electric network,
- impulse of force in time gives an instantaneous momentum to a mechanical system,
- impulse of temperature gives a unit energy, that of pressure gives a unit mass and that of concentration delivers an impulse of component mass to a process system,

- impulse of mass density in space represents a mass-point,
- impulse of charge density in space represents a point charge.

Definition 2.1.2 (Unit step function)
Integration of the unit impulse gives the unit step function

$$\eta(t) = \int_{-\infty}^{t} \delta(\tau)d\tau \tag{2.2}$$

which therefore reads as

$$\eta(t) = \begin{cases} 0 & \text{if } t < 0 \\ 1 & \text{if } t \geq 0 \end{cases} \tag{2.3}$$

Example 2.1.2 (Unit impulse and unit step signals)
Unit impulse as the derivative of the unit step

As an example of a method for dealing with generalized functions, consider the following function:

$$x(t) = \frac{d}{dt}\eta(t)$$

with η being the unit step function defined above. Since η is discontinuous, its derivative does not exist as an ordinary function, but it exists as a generalized function. Let's put x in an integral with a smooth testing function f.

$$y(t) = \int_{-\infty}^{\infty} f(t)\frac{d}{dt}\eta(t)dt$$

and calculate the integral using the integration-by-parts theorem

$$y(t) = f(t)\eta(t)\big|_{-\infty}^{\infty} - \int_{-\infty}^{\infty} \eta(t)\frac{d}{dt}f(t)dt$$

which gives

$$y(t) = f(\infty) - \int_{0}^{\infty} \frac{d}{dt}f(t)dt = f(0)$$

This results in

$$\int_{-\infty}^{\infty} f(t)\frac{d}{dt}\eta(t)dt = f(0)$$

---

which, from the defining Equation (2.1), implies that

$$\delta(t) = \frac{d}{dt}\eta(t)$$

That is, the unit impulse is the derivative of the unit step in a generalized function sense.

2.1.4 Operations on Signals

Operations on signals are used to derive (possibly more complex) signals from elementary signals or to extract some of the important signal properties.

Elementary Operations. Let ν be an n-dimensional vector space with an inner product $<\cdot,\cdot>_\nu$, $\alpha \in \mathbb{R}$ and $x, y : \mathbb{R}_0^+ \mapsto \nu$, *i.e.*

$$x(t) = \begin{bmatrix} x_1(t) \\ \vdots \\ x_n(t) \end{bmatrix}, \quad y(t) = \begin{bmatrix} y_1(t) \\ \vdots \\ y_n(t) \end{bmatrix}$$

Sum of signals. The sum of x and y is defined point-wise, *i.e.*

$$(x+y)(t) = x(t) + y(t), \quad \forall t \in \mathbb{R}_0^+ \tag{2.4}$$

Multiplication by scalar.

$$(\alpha x)(t) = \alpha x(t), \quad \forall t \in \mathbb{R}_0^+ \tag{2.5}$$

Inner product of signals. The inner product of x and y is defined as

$$\langle x, y \rangle_\nu(t) = \langle x(t), y(t) \rangle_\nu, \quad \forall t \in \mathbb{R}_0^+ \tag{2.6}$$

If the inner product on ν for $a, b \in \nu$, $a = [a_1 \; \dots \; a_n]^T$, $b = [b_1, \; \dots \; b_n]^T$ is defined as

$$\langle a, b \rangle = \sum_{i=1}^{n} a_i b_i \tag{2.7}$$

then (2.6) simply gives $\langle x, y \rangle_\nu(t) = x^T(t)y(t)$, which is a simple point-wise product if $\dim(\nu) = 1$.

Time shifting. For $a \in \mathbb{R}$, the time shifting of x is defined as

$$\mathbf{T}_a x(t) = x(t-a) \tag{2.8}$$

Causal time shifting. For $a \in \mathbb{R}$, the causal time shifting differs from ordinary time shifting in the fact that the value of the original signal before $t = 0$ is not taken into consideration, *i.e.*

$$\mathbf{T}_a^c x(t) = \eta(t - a)x(t - a) \tag{2.9}$$

where η is the unit step function defined in Equation (2.3).

Truncation. The value of a truncated signal after the truncation time T is zero, *i.e.*

$$x_T(t) = \begin{cases} x(t), 0 \le t < T \\ 0, \quad t \ge T \end{cases} \tag{2.10}$$

Convolution. Convolution is a very important binary time-domain operation in linear systems theory as we will see later (*e.g.* it can be used for computing the response of a linear system to a given input).

Definition 2.1.3 (Convolution of signals)
Let $x, y : \mathbb{R}_0^+ \mapsto \mathbb{R}$. The convolution of x and y denoted by $x * y$ is given by

$$(x * y)(t) = \int_0^t x(\tau)y(t - \tau)d\tau, \quad \forall t \ge 0 \tag{2.11}$$

Example 2.1.3 (Convolution of simple signals)

Let us compute the convolution of the signals $x, y : \mathbb{R}_0^+ \mapsto \mathbb{R}$, $x(t) = 1$, $y(t) = e^{-t}$. According to the definition

$$(x * y)(t) = \int_0^t 1 \cdot e^{-(t-\tau)}d\tau = e^{-t}\left[e^\tau\right]_0^t = 1 - e^{-t}$$

The signals and their convolution are shown in Figure 2.2.

Laplace Transformation. The main use of Laplace transformation in linear systems theory is to transform linear differential equations into algebraic ones. Moreover, the transformation allows us to interpret signals and linear systems in the frequency domain.

Definition 2.1.4 (The domain of Laplace transformation)
The domain Λ of Laplace transformation is the set of integrable complex-valued functions mapping from the set of real numbers whose absolute value is not increasing faster than exponentially, i.e.

$$\Lambda = \{ f \mid f : \mathbb{R}_0^+ \mapsto \mathbb{C}, f \text{ is integrable on } [0, a] \; \forall a > 0 \text{ and}$$
$$\exists A_f \ge 0, a_f \in \mathbb{R} \text{ such that } |f(x)| \le A_f e^{a_f x} \; \forall x \ge 0 \} \tag{2.12}$$

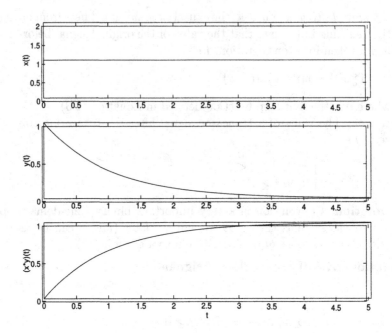

Figure 2.2. Two signals ($x(t) = 1$, $y(t) = e^{-t}$) and their convolution

Definition 2.1.5 (Laplace transformation)
With the domain defined above, the Laplace transformation of a signal f is defined as

$$\mathcal{L}\{f\}(s) = \int_0^\infty f(t)e^{-st}dt, \quad f \in \Lambda, \ s \in \mathbb{C} \tag{2.13}$$

The most important properties (among others) of Laplace transformation that we will use later are the following:

Let $f, g \in \Lambda$. Then the following equalities hold

1. *Linearity*

$$\mathcal{L}\{c \cdot f\} = c \cdot \mathcal{L}\{f\}, \quad c \in \mathbb{C} \tag{2.14}$$
$$\mathcal{L}\{f + g\} = \mathcal{L}\{f\} + \mathcal{L}\{g\} \tag{2.15}$$

2. *Laplace transform of the derivative*

$$\mathcal{L}\{f'\}(s) = s \cdot \mathcal{L}\{f\}(s) - f(0^+) \tag{2.16}$$

3. *Convolution theorem*

$$\mathcal{L}\{f * g\} = \mathcal{L}\{f\} \cdot \mathcal{L}\{g\} \tag{2.17}$$

4. *Time shifting*

$$\mathcal{L}\{\mathbf{T}_a^c(f)\} = e^{-as}\mathcal{L}\{f\}(s), \quad a \in \mathbb{R} \tag{2.18}$$

5. *Modulation*

$$\mathcal{L}\{e_\lambda f\}(s) = \mathcal{L}\{s - \lambda\} \tag{2.19}$$

where $\lambda \in \mathbb{C}$ and $e_\lambda(t) = e^{\lambda t}$.

2.1.5 L_q Signal Spaces and Signal Norms

Signals with similar mathematical properties belong to signal spaces which are usually equipped by suitable signal norms. L_q spaces are the most frequently used general signal spaces described below.

Definition 2.1.6 (L_q spaces, scalar case)
For $q = 1, 2, \ldots$ the signal space $L_q[0, \infty)$ contains the functions $f : \mathbb{R}_0^+ \mapsto \mathbb{R}$, which are Lebesgue-measurable (i.e. their generalized integral exists in the Lebesgue sense) and satisfy

$$\int_0^\infty |f(t)|^q dt \leq \infty \tag{2.20}$$

The magnitude of functions in an L_q space is measured using norms (see Section A.1 in the Appendix for the defining properties of norms).

Definition 2.1.7 (q-norm, scalar case)
Let $f \in L_q[0, \infty)$ for $q = 1, 2, \ldots$. The q-norm of f denoted by $\|f\|_q$ is defined as

$$\|f\|_q = \left(\int_0^\infty |f(t)|^q dt \right)^{\frac{1}{q}} \tag{2.21}$$

It is known that $L_q[0, \infty)$ are complete normed linear spaces (Banach spaces) with respect to the q-norms.

L_q spaces can be extended further in the following way:

Definition 2.1.8 (L_{qe} spaces, scalar case)
For $q = 1, 2, \ldots$ the signal space L_{qe} consists of the functions $f : \mathbb{R}_0^+ \mapsto \mathbb{R}$, which are Lebesgue-measurable and $f_T \in L_q$ for all $T, 0 \leq T < \infty$.

Note that the L_{qe} spaces are not normed spaces.

It is easy to see that $L_q \subset L_{qe}$. For this, consider a function $f \in L_q$. Then for any $0 < T < \infty$ $\int_0^\infty |f_T(t)|^q dt \leq \infty$ and therefore $f \in L_{qe}$. For the opposite direction, consider the signal $g(t) = \frac{1}{t+1}$ and $q = 1$. It's clear that for $T < \infty$, $\int_0^T |g(t)| dt = \ln(T+1) - \ln(1)$, but the L_1 norm of g is not finite since $\lim_{t \to \infty} \ln(t) = \infty$. Therefore $g \in L_{1e}$ but $g \notin L_1$.

For the treatment of multi-input–multi-output systems, we define L_q spaces, q-norms and L_{qe} spaces for vector-valued signals, too. For this, consider a finite dimensional normed linear space ν equipped with a norm $\| \cdot \|_\nu$.

Definition 2.1.9 (L_q spaces, vector case)
For $q = 1, 2, \ldots$ the signal space $L_q(\nu)$ contains the functions $f : \mathbb{R}_0^+ \mapsto \nu$, which are Lebesgue-measurable and satisfy

$$\int_0^\infty \|f(t)\|_\nu^q dt < \infty \tag{2.22}$$

Definition 2.1.10 (q-norm, vector case)
Let $f \in L_q(\nu)$ for $q = 1, 2, \ldots$. The q-norm of f denoted by $\|f\|_q$ is defined as

$$\|f\|_q = \left(\int_0^\infty \|f(t)\|_\nu^q dt \right)^{\frac{1}{q}} \tag{2.23}$$

Definition 2.1.11 (L_{qe} spaces, vector case)
For $q = 1, 2, \ldots$ the signal space $L_{qe}(\nu)$ consists of the functions $f : \mathbb{R}_0^+ \mapsto \nu$, which are Lebesgue-measurable and $f_T \in L_q(\nu)$ for all T, $0 \le T < \infty$.

Special Cases. Two important special cases for the q-norms are given below (compare with Subsection A.1.2).

1. $q = 2$
   For $f \in L_2$

   $$\|f\|_2 = \left(\int_0^\infty f^2(t) \right)^{\frac{1}{2}} \tag{2.24}$$

   which can be associated with the inner product

   $$\langle f, f \rangle = \int_0^\infty f^2(t) dt \tag{2.25}$$

   Similarly, in the case of $f \in L_2(\nu)$

   $$\|f\|_2 = \left(\int_0^\infty \langle f(t), f(t) \rangle_\nu dt \right)^{\frac{1}{2}} = \langle f, f \rangle^{\frac{1}{2}} \tag{2.26}$$

2. $q = \infty$
   For $f \in L_\infty$

   $$\|f\|_\infty = \sup_{t \in \mathbb{R}_0^+} |f(t)| \tag{2.27}$$

   and for $f \in L_\infty(\nu)$

   $$\|f\|_\infty = \sup_{t \in \mathbb{R}_0^+} \|f(t)\|_\nu \tag{2.28}$$

Example 2.1.4 (Signal spaces and their relations)

Besides the L_q-spaces, we can define many other signal spaces. A few examples are given below.

$\mathcal{F}(T)$ vector space: let V be the set of functions mapping from a set T to the set of real or complex numbers, where the operations are defined point-wise, *i.e.* for $x, y \in V$ and $\alpha \in \mathbb{K}$

$$(x + y)(t) = x(t) + y(t), \quad \forall t \in T$$
$$(\alpha x)(t) = \alpha \cdot x(t), \quad \forall t \in T$$

The following vector spaces are the subspaces of $\mathcal{F}(T)$:

$\mathcal{B}(T)$ vector space: $\mathcal{B}(T)$ is the set of bounded functions mapping from T to the set of real or complex numbers.

$\mathcal{C}([a, b])$ vector space: $\mathcal{C}([a, b])$ is the set of functions which are continuous on the closed interval $[a, b]$.

$\mathcal{C}^{(k)}([a, b])$ vector space: let $k \in \mathbb{N}$, $k > 1$. The set of functions which are k times continuously differentiable on the interval $[a, b]$ is a vector space and is denoted by $\mathcal{C}^{(k)}([a, b])$. At the endpoints of the interval a and b, the left and right derivatives of the functions are considered respectively.

It is clear that the following relations hold in the above examples in the case when $T = [a, b]$:

$$\mathcal{C}^{(k)}[a, b] \subset \mathcal{C}[a, b] \subset \mathcal{B}([a, b]) \subset \mathcal{F}([a, b]) \tag{2.29}$$

2.2 Systems

A system can be defined as a physical or logical device that performs an operation on a signal. Therefore we can say that systems process input signals to produce output signals.

A system is a set of objects linked by different interactions and relationships. The elements and boundaries of a system are determined by the interactions and mutual relationships, that are taken into consideration.

The nature and outcome of the interactions in physical systems is governed by certain laws. These laws can be used as *a priori* information in further examinations. The information about the system should be used in a well-defined form.

A special group of information with major importance is the one that gives the current *state* of the system. The state of a system is the collection of all

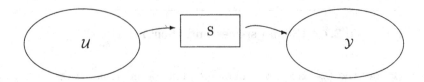

Figure 2.3. System mapping the elements of the input signal space to the output signal space

the information that describes the relations between the different interactions of the system at a given time instant (compare with Section 3.1 later). Two important kinds of information must be known in order to determine the state of the system: firstly, the *structure* of the system and secondly, the *parameters* of the system. A system can be described structurally by defining the *system topology, the rules of interconnection of the elements* and the *functional descriptions of the elements (constitutive relations)*.

Based on the above, a system can be considered as an abstract operator mapping from the input signal space to the output signal space (see Figure 2.3). The notation of this is

$$y = \mathbf{S}[u], \quad u \in \mathcal{U}, \, y \in \mathcal{Y} \tag{2.30}$$

where $\mathbf{S}$ is the system operator, u is the input, y is the output, and $\mathcal{U}$ and $\mathcal{Y}$ denote the input and output signal spaces respectively.

2.2.1 Classification of Systems: Important System Properties

Causality. The "present" in a causal system does not depend on the "future" but only on the past. This applies for every signal that belongs to the system and to the system operator $\mathbf{S}$ as well.

Definition 2.2.1 (Causal system)
A system is called causal if $y(t)$ does not depend on $u(t + \tau)$, $\forall t \geq 0, \tau > 0$.

Physical systems where time is the independent variable are causal systems. However, there are some systems that are not causal, *e.g.*

- some optical systems where the independent variables are the space coordinates,
- many off-line signal processing filters when the whole signal to be processed is previously recorded.

Linearity. A property of special interest is linearity.

Definition 2.2.2 (Linear system)
A system $\mathbf{S}$ is called linear if it responds to a linear combination of its possible input functions with the same linear combination of the corresponding output functions. Thus for the linear system we note that:

$$\mathbf{S}[c_1 u_1 + c_2 u_2] = c_1 \mathbf{S}[u_1] + c_2 \mathbf{S}[u_2] \qquad (2.31)$$

with $c_1, c_2 \in \mathbb{R}$, $u_1, u_2 \in \mathcal{U}$, $y_1, y_2 \in \mathcal{Y}$ and $\mathbf{S}[u_1] = y_1$, $\mathbf{S}[u_2] = y_2$.

Continuous Time and Discrete Time Systems. We may classify systems according to the time variable $t \in \mathcal{T}$ we apply to their description (see Subsection 2.1.2 for the definition of continuous and discrete time signals). There are continuous time systems where time is an open interval of the real line ($\mathcal{T} \subseteq \mathcal{R}$). Discrete time systems have an ordered set $\mathcal{T} = \{\cdots, t_0, t_1, t_2, \cdots\}$ as their time variable set.

SISO and MIMO Systems. Here the classification is determined by the number of input and output variables. The input and output of a single input–single output (SISO) system is a scalar value at each time instant, while multi input–multi output (MIMO) systems process and produce vector-valued signals.

Time-invariant and Time-varying Systems. The second interesting class of systems are time-invariant systems. A system $\mathbf{S}$ is *time-invariant* if its response to a given input is invariant under time shifting. Loosely speaking, time-invariant systems do not change their system properties in time. If we were to repeat an experiment under the same circumstances at some later time we get the same response as originally observed. This situation is depicted in Figure 2.4 below.

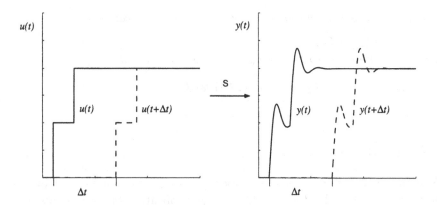

Figure 2.4. The notion of time invariance

The system parameters of a time-invariant system are constants, i.e. they do not depend on time.

We can define the notion of time invariance in a rigorous mathematical way using the shift operator defined in Equation (2.8).

Definition 2.2.3 (Time-invariant system)

A system is called time-invariant if its system operator commutes with the time shift operator, i.e.

$$T_a \circ S = S \circ T_a, \quad \forall a \in \mathbb{R} \tag{2.32}$$

2.2.2 Input–output Stability: L_q-stability and L_q-gain

As we will see later in Chapter 7, it is of interest from the viewpoint of stability how a system operator changes the norm of the input signals. This property is expressed in the L_q-stability and L_q-gain of a system.

For the following definitions, assume that $\mathcal{U}$ and $\mathcal{Y}$ are finite dimensional linear spaces of the input and output signals respectively, and S is a system operator mapping from $L_{qe}(\mathcal{U})$ to $L_{qe}(\mathcal{Y})$.

Definition 2.2.4 (L_q-stability)

S is called L_q-stable if

$$u \in L_q(\mathcal{U}) \Rightarrow G(u) \in L_q(\mathcal{Y}) \tag{2.33}$$

Definition 2.2.5 (Finite L_q-gain)

S is said to have finite L_q-gain if there exist finite constants γ_q and b_q such that

$$\|(S[u])_T\|_q \leq \gamma_q \|u_T\|_q + b_q, \quad \forall T \geq 0, \ \forall u \in L_{qe}(\mathcal{U}) \tag{2.34}$$

It is said that S has finite L_q-gain with zero bias if b_q can be zero in (2.34).

Definition 2.2.6 (L_q-gain)

Let S have finite L_q-gain. The L_q-gain of S is defined as

$$\gamma_q(S) = \inf\{\gamma_q \mid \exists b_q \text{ such that (2.34) holds}\} \tag{2.35}$$

2.3 Summary

Systems are described as abstract operators acting on signal spaces in this basic introductory chapter. The definition, classification and basic operations on both signals and systems are summarized, which will be extensively used throughout the whole book.

2.4 Questions and Exercises

Exercise 2.4.1. Give examples of signals of special importance. What is the relationship between the unit impulse and the unit step signal?

Exercise 2.4.2. What are the operations that are defined on signals? Characterize the elementary- and integral-type operations. Compute the convolution of an arbitrary signal

$$x : \mathbb{R}_0^+ \mapsto \mathbb{R}$$

with the shifted Dirac-δ function δ_τ which has its singular point at $t = \tau$.

Exercise 2.4.3. Give the most important system classes. Define the class of continuous time linear time-invariant (LTI) systems.

Exercise 2.4.4. Give the definitions of L_q and L_{qe} spaces and their underlying norms both in the scalar and vector case. Compare your definitions with the special cases of signal norms in Section A.1.2 in the Appendix.

Exercise 2.4.5. Consider the following signal:

$$f : \mathbb{R}_0^+ \mapsto \mathbb{R}, \quad f(t) = \exp(-t^2), \ t \geq 0$$

1. Show that $f \in L_1$ and $f \in L_\infty$.
2. Calculate the L_1 and L_∞ norm of f.

Exercise 2.4.6. The following two-dimensional signal is given:

$$g : \mathbb{R}_0^+ \mapsto \mathbb{R}^2, \quad f(t) = \begin{bmatrix} f_1(t) \\ f_2(t) \end{bmatrix} = \begin{bmatrix} -1 \\ 3 \cdot \exp(-2t) \end{bmatrix}, \ t \geq 0$$

1. Is $g \in L_2$?
2. Is $g \in L_{2e}$?
3. Let $\| \cdot \|_\nu$ be the normal Euclidean norm on $\mathbb{R}^2$. Calculate $\|g(t)\|_\nu$ for $t \geq 0$ and $\|g\|_\infty$.

Exercise 2.4.7. Calculate the convolution of f_1 and f_2 in Exercise 2.4.6 and check the convolution theorem for the Laplace transform of f_1 and f_2.

3. State-space Models

State-space models are the natural form of system models for a process engineer. The reason for this is that process engineering models originating from first engineering principles can be transformed into state-space form in the majority of practically important cases [32]. As a consequence, state variables in a process system model possess a clear engineering meaning – they are the canonical set of conserved extensive quantities in the process system.

> *Throughout the book we restrict ourselves to the class of finite dimensional systems, that is, to systems with finite dimensional state variables.*
>
> *In process systems engineering terminology, such systems are assumed to consist of a finite number of perfectly stirred (or mixed) balance volumes and are called lumped systems [32].*

This chapter introduces the fundamentals and basic notions of state-space representations and the most important classes of state-space models in the following sections:

- *Basic notions*
  The concept of state is introduced.
- *Linear time-invariant state-space models*
  A classical well-known state-space model class for which closed analytical solutions exist for most of the analysis and control problems [9], [39].
- *Linear time-varying and linear parameter-varying systems*
  Special system classes which are relatively easy-to-handle extensions of the classical LTI case but are capable of describing some of the essentially nonlinear system classes.
- *Nonlinear concentrated parameter state-space models*
  The input-affine form [37] is the system class this book deals with. As we shall see later in Chapter 4, majority of the lumped parameter index-1 process models can be written in this form.

3.1 Basic Notions of State-space Representation

State-space models rely on the general concept of state. If we describe a system as an operator mapping from the space of inputs to the space of outputs (see Section 2.2 for details), then we may need the entire input–output history of the system together with the planned input in order to compute future output values. This is in good agreement with the abstract, operator-based description of a system that is presented in Chapter 2.

Alternatively, we may use new information (which is called the *state of the system at t_0*) that contains all past information on the system up to time t_0 including the initial conditions for the outputs and its derivatives as well as the past input history. Then, to compute $y(t)$ for $t \geq t_0$ (all future values) we only need $u(t)$, $t \geq t_0$ and the state $x(t)$ at $t = t_0$.

The Concept of State. We may construct a signal, which is a time-dependent function, from the values of the state $x(t)$ at any given time t, which is called the state signal x. Similarly to the input and output signals, we abbreviate the state signal simply to "state".

> *A description that uses the state signal is called a state-space description or state-space model.*

State-space models for lumped (concentrated-parameter) systems consist of two sets of equations:

1. *State equations*, which describe the evolution of the states as a function of the input and state variables, being a set of time-dependent ordinary differential equations.
2. *Output equations*, which relate the value of the output signals to the state and the input signals, being algebraic equations.

We recall that any kind of system model is in fact a realization of the abstract system operator **S** (see Chapter 2). Note that a realization of an operator mapping from a space $\mathcal{U}$ to $\mathcal{Y}$ is obtained if one fixes a basis and therefore a coordinate system in $\mathcal{U}$ and $\mathcal{Y}$, and describes the mapping in these coordinates.

It is important to note that only a finite number of state equations and the same number of state variables are needed to describe lumped systems, therefore they are called *finite dimensional systems*. This corresponds to the fact that there exists a finite dimensional basis in both the input space $\mathcal{U}$ and the output space $\mathcal{Y}$ in this case.

3.2 Finite Dimensional Linear Time-invariant (LTI) Systems

The most simple and yet important system class is the class of finite dimensional linear time-invariant systems (abbreviated as FDLTI, or simply LTI systems). These systems and the analysis and control techniques based thereon are the subject of standard introductory courses on systems and control based on a state-space approach. The majority of the well-known and popular analysis and control techniques, such as controllability and observability analysis, Kalman filtering or LQR (Linear Quadratic Regulator) theory, were developed and investigated for such systems and later on have been extended to more complicated system classes.

Therefore, the LTI case will be used as a reference case throughout the book. Most of the new techniques and approaches will build on these used for LTI systems.

3.2.1 The General Form of State-space Models

For continuous time LTI systems the general form of state-space models is as follows:

$$\begin{aligned}
\dot{x}(t) &= Ax(t) + Bu(t) \qquad (\textit{state equation}) \\
y(t) &= Cx(t) + Du(t) \qquad (\textit{output equation})
\end{aligned} \tag{3.1}$$

with given initial condition $x(t_0) = x(0)$ and

$$x(t) \in \mathbb{R}^n, \ y(t) \in \mathbb{R}^p, \ u(t) \in \mathbb{R}^r$$

being the state, output and input vectors of finite dimensional spaces and

$$A \in \mathbb{R}^{n \times n}, \ B \in \mathbb{R}^{n \times r}, \ C \in \mathbb{R}^{p \times n}, \ D \in \mathbb{R}^{p \times r}$$

being matrices with constant (time-independent) elements.

It is important to note the above state-space model applies also for MIMO systems when $p > 1$, $r > 1$.

> *Most often we suppress t denoting the time dependence from the state, input and output signals and write simply x instead of x(t).*

Definition 3.2.1 (LTI state-space representation)
The state-space representation (SSR) of LTI systems is the quadruplet of constant matrices (A, B, C, D) in Equation (3.1). The dimension of an SSR is the dimension of the state vector: $\dim x(t) = n$. The state-space $\mathcal{X}$ is the set of all states:

$$x(t) \in \mathcal{X}, \quad \dim \mathcal{X} = n$$

Example 3.2.1 (LTI state-space model equation)

Let us consider the following state-space model:

$$\dot{x} = \begin{bmatrix} -\frac{v_c}{V_c} - \frac{UA}{c_{Pc}\rho_c V_c} & \frac{UA}{c_{Pc}\rho_c V_c} \\ \frac{UA}{c_{Ph}\rho_h V_h} & -\frac{v_h}{V_h} - \frac{UA}{c_{Ph}\rho_h V_h} \end{bmatrix} x + \begin{bmatrix} \frac{v_c}{V_c} & 0 \\ 0 & \frac{v_h}{V_h} \end{bmatrix} u \quad (3.2)$$

$$y = x \quad (3.3)$$

with $x, u, y \in \mathcal{R}^2$ being two-dimensional vectors. If all the model parameters

$$v_c, \ V_c, \ U, \ A, \ c_{Pc}, \ \rho_c, \ v_h, \ V_h, \ c_{Ph}, \ \rho_h$$

are constants, then the above model is indeed linear and time-invariant, with the state equation (3.2) and output equation (3.3) and with the special output matrices ($C = I$, $D = 0$).

Note that the above LTI state-space model describes a heat exchanger cell, the process system description of which is introduced in detail later in Subsection 4.4.2.

3.2.2 Linear Transformation of States

State-space models are not unique: if we have a state-space representation (A, B, C, D) for an LTI system, then we can easily find infinitely many other ones with the same dimension.

Definition 3.2.2 (Equivalent state-space models)
Two state-space representations are equivalent if they have the same input–output description.

Equivalent state-space models of LTI systems can be generated by applying coordinate transformation on the state-space. Transforming the coordinates in the state-space is often very useful in order to highlight some properties of interest (*e.g.* reachability, observability, *etc.*), or to show how certain control problems can be solved.

In the case of LTI state-space models, equivalent state-space representations can be obtained by *linear coordinate transformation*, which relates two possible equivalent state-space models

$$\begin{aligned} \dot{x}(t) &= Ax(t) + Bu(t), & \dot{\bar{x}}(t) &= \overline{A}\bar{x}(t) + \overline{B}u(t) \\ y(t) &= Cx(t) + Du(t), & y(t) &= \overline{C}\bar{x}(t) + \overline{D}u(t) \end{aligned} \quad (3.4)$$

related by the transformation

$$T \in \mathbb{R}^{n \times n}, \quad \det T \neq 0$$

i.e.

$$\overline{x} = Tx \quad \Rightarrow \quad x = T^{-1}\overline{x} \tag{3.5}$$

Observe that we do not transform the input and output signals because we want to preserve the input–output behavior of the state representations to have equivalent state-space models.

If we transform the first set of state equations in Equation (3.4) using the transformation matrix T, we get

$$\dim \mathcal{X} = \dim \overline{\mathcal{X}} = n \tag{3.6}$$

and

$$T^{-1}\dot{\overline{x}} = AT^{-1}\overline{x} + Bu$$

so finally

$$\dot{\overline{x}} = TAT^{-1}\overline{x} + TBu, \quad y = CT^{-1}\overline{x} + Du \tag{3.7}$$

In this way, we can define infinitely many state variables for the same system, and the realization matrices are related by

$$\overline{A} = TAT^{-1}, \quad \overline{B} = TB, \quad \overline{C} = CT^{-1}, \quad \overline{D} = D \tag{3.8}$$

3.2.3 Special Realization Forms of LTI Systems

Realizations of special forms play an important role in investigating the dynamic properties of an LTI system.

Diagonal Form Realization. Diagonal form realization is characterized by a diagonal state matrix A that plays a role in analyzing asymptotic stability. It is important to note that there are systems which cannot be transformed into diagonal forms (it is known from linear algebra that the necessary and sufficient condition for diagonalizing an $n \times n$ quadratic matrix is that it has n linearly independent eigenvectors).

Definition 3.2.3 (Diagonal form realization of LTI systems)
A diagonal form realization is a realization with the matrices (A, B, C) in the following special form:

$$A = \begin{bmatrix} \lambda_1 & \cdots & 0 \\ & \cdot & \cdot & \cdot & \cdot \\ \cdot & \cdot & \cdots & \cdot \\ & \cdot & \cdot & \cdot & \cdot \\ 0 & \cdots & \lambda_n \end{bmatrix}, \quad B = \begin{bmatrix} b_1 \\ \cdot \\ \cdot \\ \cdot \\ b_n \end{bmatrix},$$

$$C = \begin{bmatrix} c_1 & \cdots & c_n \end{bmatrix}$$

Note that the λ_i parameters present in the state matrix A of a diagonal form realization are the eigenvalues of the matrix and the poles of the system. A suitable transformation matrix T that brings a given realization (A, B, C) into its diagonal form can be constructed from the eigenvectors of the state matrix A.

Controller Form Realization. This realization plays a role in assessing the joint controllability and observability of an LTI system because it is always controllable.

Definition 3.2.4 (Controller form realization of LTI systems)
The controller form realization of an LTI system is given by the state-space model

$$\dot{x}(t) = A_c x(t) + B_c u(t)$$
$$y(t) = C_c x(t)$$

(3.9)

with the matrices of the following special form:

$$A_c = \begin{bmatrix} -a_1 & -a_2 & \dots & -a_n \\ 1 & 0 & \dots & 0 \\ \cdot & \cdot & \cdot & \cdot \\ \cdot & \cdot & \cdot & \cdot \\ \cdot & \cdot & \cdot & \cdot \\ 0 & 0 & \dots 1 & 0 \end{bmatrix}$$

with the coefficients of the polynomial $a(s) = s^n + a_1 s^{n-1} + \dots + a_{n-1} s + a_n$ and

$$B_c = \begin{bmatrix} 1 \\ 0 \\ \cdot \\ \cdot \\ \cdot \\ 0 \end{bmatrix}$$

$$C_c = \begin{bmatrix} b_1 & b_2 & \dots & b_n \end{bmatrix}$$

with the coefficients of the polynomial $b(s) = b_1 s^{n-1} + \dots + b_{n-1} s + b_n$ where the above polynomials appear in the transfer function $H(s) = \frac{b(s)}{a(s)}$ (see later in Definition 5.1.2 in Subsection 5.1.2).

3.3 Linear Time-varying (LTV) Parameter Systems

The simplest extension to the basic LTI case is the case of linear time-varying parameter systems (abbreviated as LTV systems) in which we allow the parameters of an LTI system to be time-varying instead of being constant. In this way we can describe certain disturbances acting as parameters in the linear model if there is information on their time variation. *In order to have a linear state-space model with time-varying parameters one needs to assume that the parameters enter into the model in a linear way.*

It is easy to see that an LTV system is still a linear system but it is no longer a time-invariant one.

In the case where the system parameters in a linear system vary with time we may generalize the state-space representation of LTI systems in the following way:

$$\begin{aligned}\dot{x}(t) &= A(t)x(t) + B(t)u(t) \quad &\text{(state equation)}\\ y(t) &= C(t)x(t) + D(t)u(t) \quad &\text{(output equation)}\end{aligned} \tag{3.10}$$

with the state, input and output vectors x, u and y and with the time-dependent realization matrices $(A(t), B(t), C(t), D(t))$.

It is important to note that the nice properties of LTI systems are partially lost or at least valid only locally for LTV systems.

Example 3.3.1 (LTV state-space model equation)

One can obtain an LTV state-space model from the simple LTI state-space model in Example 3.2.1 by assuming that the model parameters v_c and v_h vary with time. This assumption makes the state and input matrices time-varying in the form:

$$A(t) = \begin{bmatrix} -\frac{v_c(t)}{V_c} - a_{12} & a_{12} \\ a_{21} & -\frac{v_h(t)}{V_h} - a_{21} \end{bmatrix}, \quad B(t) = \begin{bmatrix} \frac{v_c(t)}{V_c} & 0 \\ 0 & \frac{v_h(t)}{V_h} \end{bmatrix} \tag{3.11}$$

with the parameters (a_{12}, a_{21}, V_c, V_h) being constants and with the special output matrices ($C = I$, $D = 0$).

Note that this is a realistic case for the modeled heat exchanger cell when we do not manipulate the flowrates v_c and v_h but consider their naturally bounded slow variation with time.

3.4 Linear Parameter-varying (LPV) Systems

A further step in the generalization of LTI systems is the case where we consider the state-space matrices of an LTI system to be fixed functions of some vector of time-varying parameters $\theta(t) \in \mathbb{R}^\ell$. These systems are called linear parameter-varying systems (abbreviated as LPV systems) and are described by state-space equations of the form:

$$\begin{aligned}\dot{x}(t) &= A(\theta(t))x(t) + B(\theta(t))u(t) \quad &\text{(state equation)}\\ y(t) &= C(\theta(t))x(t) + D(\theta(t))u(t) \quad &\text{(output equation)}\end{aligned} \tag{3.12}$$

It is clearly seen that an LTV system is a special case of an LPV system when the vector of time-varying parameters equals time, *i.e.*

$$\theta(t) = t, \ \ell = 1 \ .$$

From a practical point of view, LPV systems have at least two interesting interpretations:

- they can be viewed as linear time-invariant (LTI) systems subject to time-varying parametric uncertainty $\theta(t)$,
- they can be models of linear time-varying plants or models resulting from linearization of nonlinear plants along trajectories of the parameter θ.

It is important to note that sometimes the whole parameter vector or some of its entries are chosen to be some of the system's signals, such as some state variables being also time-varying quantities. With this approach we can describe a truly nonlinear system as an LPV system, and this may have advantages.

Of particular interest will be those LPV systems in which the system matrices *affinely depend on* θ, that is

$$\begin{aligned}
A(\theta) &= A_0 + \theta_1 A_1 + \cdots + \theta_\ell A_\ell \\
B(\theta) &= B_0 + \theta_1 B_1 + \cdots + \theta_\ell B_\ell \\
C(\theta) &= C_0 + \theta_1 C_1 + \cdots + \theta_\ell C_\ell \\
D(\theta) &= D_0 + \theta_1 D_1 + \cdots + \theta_\ell D_\ell
\end{aligned} \tag{3.13}$$

In addition, we usually assume that the underlying time-varying parameter θ varies in a convex matrix polytope Θ with corner-point vectors

$$\{\omega_1, \omega_2, \ldots, \omega_\ell\}, \quad \omega_i \in \mathbb{R}^\ell$$

such that

$$\theta(t) \in \Theta := Co\{\omega_1, \omega_2, \ldots, \omega_\ell\} := \{\sum_{i=1}^{\ell} \alpha_i \omega_i \ : \ \alpha_i \geq 0, \ \sum_{i=1}^{\ell} \alpha_i = 1\} \tag{3.14}$$

This way, we construct the parameter θ as a convex linear combination of the vectors $\{\omega_1, \omega_2, \ldots, \omega_\ell\}$. This implies that every entry in the parameter vector is bounded.

Again, special nonlinear systems, such as *bilinear systems*, can be described as affine LPV systems by considering some of the state and/or input signals as time-varying parameters.

Example 3.4.1 (LPV state-space model equation)

One can obtain n LPV state-space model from the simple LTI state-space model in Example 3.2.1 by assuming that the model parameter U, the heat transfer coefficient, varies with time. This assumption makes the state matrix A dependent on the time-varying parameter $\theta(t) = U(t)$ in the form:

$$A(U(t)) = \begin{bmatrix} -\frac{v_c}{V_c} - \frac{AU(t)}{c_{Pc}\rho_c V_c} & \frac{AU(t)}{c_{Pc}\rho_c V_c} \\ \frac{AU(t)}{c_{Ph}\rho_h V_h} & -\frac{v_h}{V_h} - \frac{AU(t)}{c_{Ph}\rho_h V_h} \end{bmatrix} \qquad (3.15)$$

The constant matrices A_0 and A_1 in the affine dependence equation (3.13)

$$A(U) = A_0 + U A_1$$

are as follows:

$$A_0 = \begin{bmatrix} -\frac{v_c}{V_c} & 0 \\ 0 & -\frac{v_h}{V_h} \end{bmatrix}, \quad A_1 = \begin{bmatrix} -\frac{A}{c_{Pc}\rho_c V_c} & \frac{A}{c_{Pc}\rho_c V_c} \\ \frac{A}{c_{Ph}\rho_h V_h} & -\frac{A}{c_{Ph}\rho_h V_h} \end{bmatrix}, \qquad (3.16)$$

assuming all the remaining parameters

$$v_c, \ V_c, \ A, \ c_{Pc}, \ \rho_c, \ v_h, \ V_h, \ c_{Ph}, \ \rho_h$$

to be constants.

The remaining three matrices in the state-space representation are constant matrices, that is

$$B = \begin{bmatrix} \frac{v_c}{V_c} & 0 \\ 0 & \frac{v_h}{V_h} \end{bmatrix} \qquad (3.17)$$

and the special output matrices are ($C = I$, $D = 0$).
Note that this is a realistic case for the modeled heat exchanger cell when we consider the natural deterioration of the heat exchanger surface due to $CaCO_3$ formation.

3.5 Nonlinear Systems

Nonlinear finite dimensional systems represent a wide class of nonlinear systems. Lumped process models derived from first engineering principles belong to this class.

The general form of state-space models of finite dimensional nonlinear systems is:

$$\dot{x}(t) = \widetilde{f}(x(t), u(t)) \qquad (state\ equation)$$
$$y(t) = \widetilde{h}(x(t), u(t)) \qquad (output\ equation)$$

(3.18)

with the state, input and output vectors x, u and y and with the smooth nonlinear mappings

$$\widetilde{f} : \mathbb{R}^n \times \mathbb{R}^r \mapsto \mathbb{R}^n, \quad \widetilde{h} : \mathbb{R}^n \times \mathbb{R}^r \mapsto \mathbb{R}^p .$$

3.5.1 The General Form of State-space Models for Input-affine Systems

If the nonlinear functions $\widetilde{f}$ and $\widetilde{h}$ above are in a special form, the so-called *input-affine form* is obtained.

Definition 3.5.1 (Input-affine nonlinear state-space models)

$$\dot{x}(t) = f(x(t)) + \sum_{i=1}^{m} g_i(x(t))u_i(t) \qquad (state\ equation)$$
$$y(t) = h(x(t)) \qquad (output\ equation)$$

(3.19)

with the same state, input and output vectors x, u and y as above, and with the smooth nonlinear mappings

$$f : \mathbb{R}^n \mapsto \mathbb{R}^p, \quad g_i : \mathbb{R}^n \mapsto \mathbb{R}, \quad h : \mathbb{R}^n \mapsto \mathbb{R}^p$$

It is important to observe that the *input signals enter into the input-affine nonlinear state-space model in a linear way*, that is, the mapping $\widetilde{f}$ in the original general nonlinear state-space model (3.18) is linear with respect to u.

3.5.2 Nonlinear Transformation of States

Transforming the coordinates in the state-space is also useful in the nonlinear case, in order to investigate dynamic properties of interest (*e.g.* reachability, observability, *etc.*) or to solve certain control problems. Of course, nonlinear coordinate transformations are usually applied to nonlinear state-space models.

Definition 3.5.2 (Nonlinear coordinate transformation)
A nonlinear change of coordinates is written as

$$z = \Phi(x)$$

(3.20)

where Φ represents an $\mathbb{R}^n$-valued function of n variables, i.e.

$$\Phi(x) = \begin{bmatrix} \phi_1(x) \\ \phi_2(x) \\ \dots \\ \phi_n(x) \end{bmatrix} = \begin{bmatrix} \phi_1(x_1, \dots, x_n) \\ \phi_2(x_1, \dots, x_n) \\ \dots \\ \phi_n(x_1, \dots, x_n) \end{bmatrix}$$

(3.21)

with the following properties:

1. Φ is invertible, i.e. there exists a function Φ^{-1} such that $\Phi^{-1}(\Phi(x)) = x$ for all x in $\mathbb{R}^n$.
2. Φ and Φ^{-1} are both smooth mappings, i.e. they have continuous partial derivatives of any order.

A transformation of this type is called a global diffeomorphism on $\mathbb{R}^n$.

Example 3.5.1 (A simple nonlinear state transformation)

Consider a simple nonlinear vector-vector function f

$$f(x_1, x_2) = f(x) = \begin{bmatrix} x_1^2 + x_2^2 \\ x_1^3 + x_2^3 \end{bmatrix}$$

and let the function Φ generating the coordinate transformation be given as

$$z = \Phi(x) = \begin{bmatrix} \sqrt{x_1} \\ \sqrt{x_2} \end{bmatrix}$$

which is invertible on the domain $x_1 \geq 0$, $x_2 \geq 0$. Its inverse function is:

$$x = \Phi^{-1}(z) = \begin{bmatrix} z_1^2 \\ z_2^2 \end{bmatrix}$$

Therefore the original function f in the transformed coordinate system will be:

$$f(z) = f(\Phi^{-1}(z)) = \begin{bmatrix} z_1^4 + z_2^4 \\ z_1^6 + z_2^6 \end{bmatrix}$$

Sometimes, a transformation having both of these properties and being defined for all x is difficult to find. Thus, in most cases, one looks rather at transformations defined only in a neighborhood of a given point the existence of which is guaranteed by the inverse function theorem. A transformation of this type is called a *local diffeomorphism*. In order to check whether a given transformation is a local diffeomorphism or not, the following result is very useful.

Suppose Φ is a smooth function defined on some subset U of $\mathbb{R}^n$. Suppose the Jacobian matrix of Φ is nonsingular at a point $x = x^0$, then, on a suitable open subset U^0 of U, containing x^0, Φ defines a local diffeomorphism.

3.5.3 Bilinear State-space Models

An important special case of general nonlinear state-space models (3.18) are the so-called *bilinear state-space models*. In these models, the nonlinear functions $\tilde{f}$ and $\tilde{h}$ become

$$\dot{x}_\ell(t) = \sum_{j=1}^{n} a_{\ell j}^{(0)} x_j(t) + \sum_{j=1}^{m} b_{\ell j}^{(0)} u_j(t)$$

$$+ \sum_{j=1}^{m} \sum_{i=1}^{n} b_{ij}^{(\ell)} x_i(t) u_j(t)$$

$$i = 1, \ldots, n \quad (state\ equations)$$

$$y_k(t) = \sum_{j=1}^{n} c_{kj}^{(0)} x_j(t) \tag{3.22}$$

$$k = 1, \ldots, p \quad (output\ equation)$$

with the same state, input and output vectors x, u and y as above.

It is easy to see that bilinear state-space models are special cases of the input-affine models, because the input signal enters into the model in a linear way.

We shall see later in Section 4.3.1 that a wide class of process systems models belong to the class of bilinear state-space models.

Example 3.5.2 (Bilinear state-space model equation)

Let us consider a modified version of the simple state-space model in Equation 3.2 by changing the input variables to have

$$u = [\ v_c,\ v_h\ ]^T$$

and considering the elements of the old input vector

$$u_O = [\ u_{O1},\ u_{O2}\ ]^T = [\ T_{cIN},\ T_{hIN}\ ]^T$$

to be constants. The following bilinear state-space model is obtained:

$$\dot{x} = \begin{bmatrix} -\frac{UA}{c_{Pc}\rho_c V_c} & \frac{UA}{c_{Pc}\rho_c V_c} \\ \frac{UA}{c_{Ph}\rho_h V_h} & -\frac{UA}{c_{Ph}\rho_h V_h} \end{bmatrix} x + \begin{bmatrix} \frac{T_{cIN}}{V_c} & 0 \\ 0 & \frac{T_{hIN}}{V_h} \end{bmatrix} u$$

$$+ \begin{bmatrix} -\frac{1}{V_c} & 0 \\ 0 & 0 \end{bmatrix} u_1 \cdot x + \begin{bmatrix} 0 & 0 \\ 0 & -\frac{1}{V_h} \end{bmatrix} u_2 \cdot x \qquad (3.23)$$

$$y = x \qquad (3.24)$$

with $x, u, y \in \mathcal{R}^2$ being two-dimensional vectors.
If all the model parameters

$$T_{cIN}, \ V_c, \ U, \ A, \ c_{Pc}, \ \rho_c, \ T_{hIN}, \ V_h, \ c_{Ph}, \ \rho_h$$

are constants, then the above model is indeed bilinear and time-invariant, with the state equation (3.23), with a trivial output equation and with the special model matrices:

$$A^{(0)} = \begin{bmatrix} -\frac{UA}{c_{Pc}\rho_c V_c} & \frac{UA}{c_{Pc}\rho_c V_c} \\ \frac{UA}{c_{Ph}\rho_h V_h} & -\frac{UA}{c_{Ph}\rho_h V_h} \end{bmatrix}, \ B^{(0)} = \begin{bmatrix} \frac{T_{cIN}}{V_c} & 0 \\ 0 & \frac{T_{hIN}}{V_h} \end{bmatrix}, \ C^{(0)} = I$$

$$B^{(1)} = \begin{bmatrix} -\frac{1}{V_c} & 0 \\ 0 & 0 \end{bmatrix}, \ B^{(2)} = \begin{bmatrix} 0 & 0 \\ 0 & -\frac{1}{V_h} \end{bmatrix}$$

All the remaining matrices are zero matrices.

3.6 Summary

The state-space representations of the most important finite dimensional (lumped) system classes are briefly reviewed and described in this chapter. They will be used in the rest of the book.

The linear time-invariant (LTI) state-space model is introduced first, since it serves as a basic case for all extensions towards the nonlinear state-space representation forms. The linear time-varying parameter (LTV) and linear parameter-varying (LPV) systems are seen as intermediate cases between the basic reference LTI and the general nonlinear cases.

The general and input-affine state-space models, introduced last, are the basis of all the nonlinear analysis and control techniques described in this book, are also introduced here.

The linear and nonlinear coordinate transformations, which generate classes of equivalent linear and nonlinear state-space models, complement the chapter.

3.7 Questions and Exercises

Exercise 3.7.1. Show that (3.1) is indeed a linear and time-invariant model.

Exercise 3.7.2. Show that an LTV system is a linear system, but it is not a time-invariant one.

Exercise 3.7.3. Prove that the state-space representations given by the matrices

1. A, B, C
2. $\bar{A} = TAT^{-1}, \bar{B} = TB, \bar{C} = CT^{-1}$

give the same transfer function.

Exercise 3.7.4. Show that the quantities

$$CA^i B, \quad i = 0, 1, \dots$$

called Markov parameters of an LTI system with state-space representation matrices (A, B, C) are invariant under linear state transformations.

Exercise 3.7.5. Consider the following functions and points in the state-space

1. $f(x) = x^2 + 3$, $x_1 = 0$, $x_2 = 3$
2. $f(x) = (x - 4)^3$, $x_1 = 4$, $x_2 = 0$

Is there any open neighborhood of x_1 and x_2 where f is invertible?

Exercise 3.7.6. The following vector-field is given:

$$f(x, y) = \begin{bmatrix} 5x + 3y \\ (x + 2)^2 \end{bmatrix} \tag{3.25}$$

Is there any open and dense subset of $\mathbb{R}^2$ where f can be inverted? If yes, give such a set and calculate the corresponding inverse function.

Exercise 3.7.7. Consider a simple nonlinear vector-vector function f

$$f(x_1, x_2) = f(x) = \begin{bmatrix} x_1^2 + x_2^2 \\ x_1^4 - x_2^4 \end{bmatrix}$$

and let the function Φ generating the nonlinear coordinate transformation be given as

$$z = \Phi(x) = \begin{bmatrix} \arcsin x_1 \\ \arcsin x_2 \end{bmatrix}$$

Apply the coordinate transformation to the function f above and give its form in the transformed coordinates.

Exercise 3.7.8. Consider the state equation of a simple two-dimensional quasi-polynomial system

$$\frac{dx_1}{dt} = x_1^{1/2} x_2^{3/2} - x_1 x_2^3 + x_1^{1/2} u_1$$

$$\frac{dx_2}{dt} = 3x_1 x_2^{1/2} - 2x_1^2 x_2$$

Apply the following quasi-polynomial nonlinear coordinate transformation

$$z = \Phi(x) = \begin{bmatrix} x_1^3 x_2^{1/3} \\ x_1^{1/4} x_2^2 \end{bmatrix}$$

to the system.

Is the resulting system model quasi-polynomial again? Why?

4. Dynamic Process Models

Dynamic process models and their properties form the background of any process control activity including model analysis, model parameter and structure estimation, diagnosis, regulation or optimal control. Therefore this chapter is entirely devoted to dynamic process modeling for control purposes: the construction and properties of lumped dynamic models. Process modeling is an important and independent area in itself within process systems engineering with good and available textbooks (see, *e.g.* [32] providing an overview and introduction to the key concepts, methods and procedures). This chapter is not intended to replace a process modeling textbook in a short form, but rather focuses on the particular characteristics of a dynamic lumped parameter process model, which is seen as a finite dimensional system model given in its state-space model form. The chapter is broken down into the following sections.

- *Process modeling for control*
  We start with a brief overview of the way process models are generally constructed together with a description of the ingredients of lumped dynamic process models.
- *State-space models of process systems*
  The special properties of finite dimensional state-space models derived from conservation balances are then described together with their system variables. A decomposition of the state equations driven by the mechanisms taking place in the process system is also given.
- *Special nonlinear process systems*
  A separate section is devoted to special finite dimensional nonlinear process systems, including bilinear systems with no source term in their conservation balance equations, and system models in differential-algebraic equation (DAE) form.
- *Examples*
  Finally three sets of simple process models giving rise to finite dimensional nonlinear state-space models are given in separate sections: heat exchanger models, continuously stirred tank reactor (CSTR) models and a model of a gas turbine. These models will be used later in the book for illustrating nonlinear analysis and control techniques.

4.1 Process Modeling for Control Purposes

A modeling task is specified by giving the description of the process system to be modeled together with the modeling goal, *i.e.* the intended use of the model. The modeling goal largely determines the model, its variables, spatial and time characteristics, as well as its resolution or level of detail and precision. Process control as a modeling goal does not require very accurate models, we only aim at about 5 percent precision in values, but process control requires dynamic models that capture the time characteristics (dead time and time constants) of the model well. Moreover, we usually use lumped parameter dynamic process models for control purposes because the resulting finite dimensional system models are much easier to handle. Therefore we usually use some kind of lumping, most often the so-called "method of lines" procedure (see, *e.g.* [32]) to obtain a lumped parameter approximation from a distributed parameter system model.

Definition 4.1.1 (Balance volume)
Parts of a lumped process system which

- *contain only one phase or pseudo-phase,*
- *can be assumed to be perfectly mixed,*

will be termed balance volumes or lumps.

Balance volumes are the elementary dynamic units of a lumped process system for which dynamic balances can be constructed and assumptions can be made.

4.1.1 General Modeling Assumptions

We restrict ourselves to the following class of systems and system models throughout the book:

> 1. Only lumped process models that result in a model in ordinary differential-algebraic equation (DAE) form are considered.
> 2. We only treat initial value problems.
> 3. All physical properties in each phase are assumed to be functions of the *thermodynamic state variables* (temperature, pressure and compositions) of that phase only.

These general assumptions ensure that we always have an index 1 model with the possibility of substituting the algebraic constitutive equations into the differential ones. This is always possible if we choose the free variables and parameters of the model, the so-called *specification*, in a proper way.

In order to get relatively simple models where the algebraic equations can all be substituted into the differential ones, we use the following two additional general assumptions in the majority of our examples:

> 4. Constant pressure is assumed in the whole process system.
>
> 5. All physical properties in each phase are assumed to be constant.

Note that the last general assumption overrides assumption 3 by stating that all physical properties do not depend on any other variable, but that they are constant.

4.1.2 The Principal Mechanisms in Process Systems

Mechanisms in a process system describe different means of material or energy transport or transformation. Therefore we encounter a great variety of possible mechanisms in a process system. In order to complete the steps of constructing a mathematical model of a process system (see Section 4.1.4 for more details) we need to analyze the modeling problem statement and decide which mechanisms should be included in the model. For lumped parameter process systems these mechanisms include, but are not limited to, the following:

1. *Convection:*
   A material and energy transport mechanism in which the conserved extensive quantities (overall mass, component masses and energy) are carried by the transport of the overall mass, *i.e.* by flows. The inflows and outflows of the balance volumes can be regarded as convection in lumped process systems.

2. *Transfer:*
   A component mass or energy transport mechanism between two phases in contact when there are no convective flows involved. The driving force for transfer is the difference between the thermodynamical state variables (temperature, pressure and compositions) in the two phases.

3. *Chemical reaction:*
   A component mass transformation mechanism, which generates the products of a chemical reaction from the reactants. It usually also involves enthalpy (energy) transformation: generation or consumption.

4. *Phase changes:*
   A phase transformation mechanism, such as evaporation, condensation, melting, boiling, crystallization, *etc.* where the chemical composition remains unchanged. It also involves enthalpy (energy) transformation.

A basic property of the mechanisms of a process system is that they are assumed to be *strictly additive*, that is, they give rise to additive terms in the conservation balance equations of a process system.

4.1.3 The Basic Ingredients of Lumped Process Models

The equations of a particular model satisfying the general modeling assumptions" in Section 4.1.1 are of two types:

- differential equations (explicit first-order nonlinear ODEs with initial conditions),
- algebraic equations.

The *differential equations* originate from conservation balances, therefore they can be termed *conservation balance equations*. The algebraic equations are usually of mixed origin: they will be called *constitutive equations*. Along with the above equations we have other model elements associated with them such as:

- modeling assumptions,
- variables and parameters,
- initial conditions,
- data (specification) of model parameters and constants.

Variables are time-varying or time-dependent quantities in process model equations. They are also called signals in system theoretical terminology. A variable x is called *differential* if its time derivative $\left(\frac{dx}{dt}\right)$ is explicitly present in the DAE model. A variable is termed *algebraic* if it is not differential.

Parameters, on the other hand, are quantities which are either constant or are regarded to be constant in a particular process model.

4.1.4 The Model Construction Procedure

In order to construct a process model satisfying a given modeling goal, the problem statement of the modeling, that is, the process system description together with the goal, should be carefully analyzed first to find the relevant mechanisms together with a suitable level of detail. These drive the construction of the process model equations, which is carried out in steps forming the model construction procedure. Good modeling practice requires a *systematic way of developing the model equations of a process system for a given purpose.* Although this procedure is usually cyclic, in which one often returns back to a previous step, the systematic procedure can be regarded as a sequence of modeling steps as follows:

Step 0. System and subsystem boundary and balance volume definitions
The outcome of this step is the set of balance volumes for mass, energy and momentum. These are the conserved extensive quantities normally considered in process systems. Moreover, the number of components is also fixed for each mass balance volume.

Step 1. Establish the balance equations
Here we set up conservation balances for mass, energy and momentum for each balance volume.

Step 2. Transfer and reaction rate specifications
The transfer rate expressions between different balance volumes in the conservation balances are specified here usually as functions of intensive quantities. The reaction rates within balance volumes are also specified.

Step 3. Property relation specifications
Mostly algebraic relationships expressing thermodynamic knowledge, such as equations of state and the dependence of physico-chemical properties on thermodynamic state variables, are considered here.

Step 4. Balance volume relation specifications
Equipment with a fixed physical volume is often divided into several balance volumes if multiple phases are present. A balance volume relation describes a relation between balance volumes and physical volumes.

Step 5. Equipment and control constraint specifications
There is inevitably the need to define constraints on process systems. These are typically in the form of equipment-operating constraints (in terms of temperatures, pressures, *etc.*) and in terms of control constraints, which define relations between manipulated and controlled variables in the system.

Step 6. Selection of design variables
The selection of design variables is highly dependent on the *application area* or *problem* and is not necessarily *process- specific*. The process itself only provides constraints on which variables are potentially relevant. The selection of design variables may greatly influence the mathematical properties of the model equations, such as the differential index.

4.1.5 Conserved Extensive and Intensive Potential Variables

Any variable characterizing a process system can be classified as either *extensive* or *intensive* depending on how this variable behaves when joining two process systems together.

Definition 4.1.2 (Extensive variable)
A variable which is proportional to the overall mass of the system, that is, which is strictly additive when joining two process systems, is termed an extensive variable.

Dictated by the basic principles of thermodynamics, there is a *canonical set of extensive variables* which is necessary and sufficient to describe uniquely a single phase process system. This set includes *overall mass, component masses and energy* for a perfectly stirred (lumped) balance volume. It is important to note that these extensive variables are conserved, therefore conservation balances can be constructed for each of them (see later in Section 4.1.6).

Potentials or *intensive quantities* are related to any of the above extensive conserved quantities.

Definition 4.1.3 (Intensive variable (potential))
Intensive variable (or potential) difference (both in space and between phases in mutual contact) causes transport (transfer or diffusion depending on the circumstances) of the related extensive quantity.

The following intensive variable (potential) – extensive variable pairs are normally considered in process systems:

- temperature to internal energy or enthalpy,
- chemical potential, or simply concentration, to mass of a component,
- pressure to overall mass (not relevant for our case due to the constant pressure general assumption in Section 4.1.1).

4.1.6 Conservation Balances

Conservation balances can be set up for any conserved extensive variable in any balance volume of a process system. Recall that overall mass, component masses and internal energy form the canonical set of conserved extensive quantities. If we consider an open balance volume with in- and outflows (convection), transport (an inter-phase mechanism) and other intra-phase transformation mechanisms, the verbal form of a conservation balance equation is as follows:

$$\left\{ \begin{array}{c} net\ change\ in \\ ext.\ quantity \end{array} \right\} = \left\{ \begin{array}{c} in\text{-} \\ flows \end{array} \right\} - \left\{ \begin{array}{c} out\text{-} \\ flows \end{array} \right\} + \left\{ \begin{array}{c} generation \\ consumption \end{array} \right\} \quad (4.1)$$

Note that there is no source term for the overall mass balance because of the mass conservation principle. Chemical reaction appears in the generation-consumption term of the component mass and the energy balances, while phase transition gives rise to a generation-consumption term in the energy balance. Inter-phase transfer also appears in the generation-consumption term of both the component mass and energy balance equations. There is no overall mass transfer between the phases because of the constant pressure assumption.

The basic equation which drives all the other conservation balances is the *overall mass balance* of the perfectly stirred balance volume j:

$$\frac{dm^{(j)}}{dt} = v_{in}^{(j)} - v_{out}^{(j)} \quad (4.2)$$

where $v_{in}^{(j)}$ and $v_{out}^{(j)}$ are the mass in- and outflow rates respectively.

Under the above conditions the general form of a differential balance equation of a conserved extensive quantity ϕ for a perfectly stirred balance volume j takes the form:

$$\frac{d\phi^{(j)}}{dt} = v_{in}^{(j)}\phi_{in}^{(j)} - v_{out}^{(j)}\phi_{out}^{(j)} + q_{\phi,transfer}^{(j)} + q_{\phi,source}^{(j)} \tag{4.3}$$

Observe that the overall mass balance (4.2) is a special case of the general balance equation (4.3) with

$$q_{m,transfer}^{(j)} = 0, \quad q_{m,source}^{(j)} = 0 \tag{4.4}$$

Note that the conserved extensive quantity $\phi^{(j)}$ of balance volume j can be any variable from the following set:

$$\phi \in \{E, (m_k, k = 1, \ldots, K)\} \tag{4.5}$$

where E is the energy, m_k is the component mass of the k-th component, with K being the number of components in the balance volume. The related intensive variable (potential) is taken from the set:

$$\Phi \in \{T, (c_k, k = 1, \ldots, K)\} \tag{4.6}$$

where T is the temperature and c_k is the concentration of the k-th component.

4.1.7 Constitutive Equations

Some of the terms in the general conservation balance equations (4.3) above call for additional algebraic equations to complete the model in order to have it in a closed solvable form. These complementary algebraic equations are called *constitutive equations*. Constitutive equations describe

- extensive–intensive relationships,
- transfer rate equations for mass transfer and heat/energy transfer,
- reaction rates,
- property relations: thermodynamical constraints and relations, such as the dependence of thermodynamical properties on the thermodynamical state variables (temperature, pressure and compositions), equilibrium relations and state equations,
- balance volume relations: relationships between the defined mass and energy balance volumes,
- equipment and control constraints.

Extensive–intensive Relationships. Recall that potentials, being intensive variables, are related to their extensive variable counterparts through algebraic equations. These involve physico-chemical properties, which may depend on other potentials or on the differential variables, because the thermodynamical state variables, temperature, pressure and concentrations are intensive quantities. An example is the well-known intensive–extensive relationship of the U–T pair

$$U = mc_V T$$

where U is the internal energy, m is the total mass, T is the temperature and c_V is the specific heat capacity (a physico-chemical property) of the material in a balance volume. Note that because of the above properties of the intensive quantities and their relation to the extensive conserved quantities, the potential form of the model equations is usually derived by additional assumptions on the physical properties. In the above example

$$c_V = \mathbf{c_V}(T, P, (c_i, i = 1, ..., K))$$

but because of the "general modeling assumptions" in Section 4.1.1 $c_V = const$ is assumed.

Transfer Rate Equations. These are used to describe the algebraic form of the transfer term $q_{\phi,transfer}^{(j)}$ in the general balance equation (4.3). Dictated by the Onsager relationship from non-equilibrium thermodynamics, this term has the following general linear form:

$$q_{\phi,transfer}^{(j)} = K_{\phi,transfer}^{(j,k)} \left(\Phi^{(j)} - \Phi^{(k)} \right) \tag{4.7}$$

Here the transfer coefficients $K_{\phi,transfer}^{(j,k)}$ are generally assumed to be constants and the driving force for the transfer between balance volumes j and k is the difference of the potential variables $\Phi^{(j)}$ and $\Phi^{(k)}$.

4.2 State-space Models of Process Systems

The state-space model of a lumped process system obeying the "general modeling assumptions" in Section 4.1.1 can be obtained by substituting the algebraic constitutive equations into the conservation balance equations. This fact is used in this section to highlight the special structural properties which characterize a nonlinear state-space model derived from conservation balance equations.

4.2.1 System Variables

A possible *state vector* $\hat{x}$ for the nonlinear state-space model is:

$$\hat{x} = [\left(E^{(j)}, (m_k^{(j)}, k = 1, ..., K) \right), j = 1, ..., C]^T \tag{4.8}$$

$$\dim(\hat{x}) = n = (K + 1) \cdot C$$

with K being the number of components and C being the number of balance volumes. Note that the linear relationship

$$m^{(j)} = \sum_{k=1}^{K} m_k^{(j)} \tag{4.9}$$

enables us to choose the complete set of component masses to be present in the state vector.

The *potential input variables* (including both manipulable input variables and disturbances) are also fixed by the state equations (4.3). They are the time-dependent variables (signals) appearing on the right-hand side of the equations, and not being state variables, that is

$$\hat{u} = [(v^{(j)}, \phi_{in}^{(j)} v^{(j)}), \ j = 1, \ldots, C]^T \tag{4.10}$$

Note that we have formed a composite input vector $(\phi_{in}^{(j)} v^{(j)})$ from two signals so that the corresponding term in the balance equation (4.3) has a homogeneous form. Any external signal present in the source relations $q_{\phi,source}^{(j)}$ should also be included in the set of potential input variables.

4.2.2 State Equations in Input-affine Form

In order to transform the general form of conservation balance equations into canonical nonlinear state equation form, the state and input variables above need to be centered using an arbitrary steady state as reference. Note that this step is not needed if there is no source term or if all sources are homogeneous functions. There can be process systems with no steady state at all, such as batch or fed-batch processes. Their variables cannot be centered, therefore their conservation balance equations cannot be transformed into a nonlinear state-space model in input-affine form.

> *For the reference input–state pair (x^*, u^*) the left-hand sides of the balance equations (4.3) are zero.*

Definition 4.2.1 (Centered variable)
A centered variable is then the difference between its actual and reference value, that is

$$\overline{\varphi} = \varphi - \varphi^*$$

The centered state variables and the centered input variables are then as follows:

$$x = [\left(\overline{E}^{(j)}, (\overline{m}_k^{(j)}, k = 1, \ldots, K)\right), \ j = 1, \ldots, C]^T, \quad |x| = n = (K+1) * C \tag{4.11}$$

$$u = [(\overline{v}^{(j)}, \overline{\phi_{in}^{(j)} v^{(j)}}), \ j = 1, \ldots, C]^T \tag{4.12}$$

With the centered state (4.11) and input vectors (4.12) the general form of lumped dynamic models of process systems can be transformed into the standard input-affine form of nonlinear concentrated parameter state-space equations:

$$\dot{x} = f(x) + \sum_{i=1}^{m} g_i(x)u_i, \qquad u \in \mathbb{R}^m, \quad f(0) = 0 \tag{4.13}$$

where $x = [x_1, \ldots, x_n]^T$ are local coordinates of a state-space manifold $\mathcal{M}$. It is important to note that *the state equations of process systems are always in an input-affine form above, because of the structure of the general state equation (4.3), if the system possesses a steady state.*

4.2.3 Decomposition of the State Equations Driven by Mechanisms

The state equations in a nonlinear input-affine state-space model of a process system are derived from the general conservation balance equation (4.3). This equation has four terms in its right-hand side, corresponding to the four principal mechanisms we take into account when constructing process models:

- input convection (inflow term),
- output convection (outflow term),
- (inter-phase) transfer,
- sources including both generation and consumption.

In addition, the concrete mathematical form of the state equations depends on the selection of the actual input variables from the set of potential ones. Let us assume that:

1. The flow rates $v^{(j)}$ (and the conserved extensive quantities at the inlet) are the input variables (manipulable input variables or disturbances).
2. There are no external sources to the system to be included in the set of potential input variables, that is

$$q_{\phi,source}^{(j)} = Q_{\phi}^{(j)}(T^{(j)}, c_1^{(j)}, \ldots, c_{K-1}^{(j)}) \tag{4.14}$$

where $Q_{\phi}^{(j)}$ is a given nonlinear function.

It can be shown that, under the above conditions, one can decompose the nonlinear vector–vector functions $f(x)$ and $g(x)$ in the nonlinear state equation (4.13) into structurally different additive parts with clear engineering meaning:

$$\dot{x} = A_{transfer}x + Q_{\phi}(x) + \sum_{i=1}^{m} N_i x u_i + B_{conv}u \tag{4.15}$$

The first term in the above equation originates from the transfer, the second from the sources, while the last two correspond to the output and input convection respectively. The coefficient matrices $A_{transfer}$, B_{conv} and $(N_i, i = 1, \ldots, m)$ are constant matrices, with $A_{transfer}$ depending on the non-negative transfer coefficients $K_{\phi,transfer}^{(j,k)}$ in Equation (4.7), while N_i is a matrix with non-negative elements that depends on the connections between the balance volumes.

It is important to note that N_i is a matrix where only its i-th column is different from zero. More on the connection matrix N_i will follow in Section 4.2.4.

It is easy to see from relation (4.7) that the linear constant (i.e. time-invariant) state matrix $A_{transfer}$ is always negative semi-definite and also has zero eigenvalues. Moreover, the nonlinear source function $Q_\phi(x)$ is of block diagonal form with the blocks joining the state variables belonging to the same balance volume.

> Thus, the decomposed state equation contains a linear state term
> for the transfer, a general nonlinear state term for the sources, a
> bilinear input term for the output, and a linear input term for the
> input convection respectively.

4.2.4 Balance Volumes Coupled by Convection

Until now we have only taken into account that the balance volumes are coupled by the transfer terms (4.7), giving rise to the linear constant (i.e. time-invariant) state matrix $A_{transfer}$. Now we consider the effect of convective flows joining balance volumes to find the form of the input and output convection matrices $(N_i, i = 1, \ldots, m)$ and B_{conv}. In order to describe the general case let us assume that the outlet flow of the balance volume j is divided into parts and fed into other balance volumes, giving rise to the equation:

$$\sum_{\ell=0}^{C} \alpha_\ell^{(j)} = 1, \quad j = 0, \ldots, C \tag{4.16}$$

where $\alpha_j^{(\ell)}$ is the fraction of the total outlet flow $v^{(\ell)}$ of balance volume ℓ that flows into balance volume j. The mass inflow of balance volume j then consists of the outflows from all the connected balance volumes, including the balance volume itself, together with a flow from the environment which is described as a pseudo-balance volume with index 0:

$$v_{in}^{(j)} = \sum_{\ell=0}^{C} \alpha_j^{(\ell)} v_{out}^{(\ell)} \quad, \quad j = 0, \ldots, C \tag{4.17}$$

Finally we collect the ratios above into a convection matrix C_{conv} as follows:

$$C_{conv} = \begin{bmatrix} -(1 - \alpha_1^{(1)}) & \alpha_1^{(2)} & \cdots & \alpha_1^{(C)} \\ \cdots & \cdots & \cdots & \cdots \\ \alpha_C^{(1)} & \alpha_C^{(2)} & \cdots & -(1 - \alpha_C^{(C)}) \end{bmatrix} \qquad (4.18)$$

Observe that only the ratios belonging to the internal (that is *not* environmental) flows are collected in the matrix above. It is important to note that because of Equation (4.16) the convection matrix C_{conv} is a column conservation matrix and therefore it is a stability matrix (see Section 7.4.2 in Chapter 7).

From theoretical and practical viewpoints there are two cases of special interest:

- free output convection,
- passive (controlled) mass convection.

These are described below and will be used throughout the book.

Free Mass Convection Network. The first special case of interest is when the output mass flow of any of the balance volumes is proportional to the overall mass in the balance volume, that is

$$v_{out}^{(j)} = \kappa^{(j)} m^{(j)}, \quad \kappa^{(j)} > 0 \qquad (4.19)$$

The inlet mass flow of the balance volume j can then be written as

$$v_{in}^{(j)} = \sum_{\ell=1}^{C} \alpha_j^{(\ell)} \kappa^{(\ell)} m^{(\ell)} + \alpha_j^{(0)} v_{out}^{(0)}, \quad j = 0, \ldots, C \qquad (4.20)$$

where $\alpha_j^{(0)} v_{out}^{(0)} = v_{IN}^{(j)}$ is the mass inflow of the balance volume from the environment, that is, the real inflow. With the free convection equation above, the overall mass balance of the balance volume j (4.2) takes the following form:

$$\frac{dm^{(j)}}{dt} = \sum_{\ell=1}^{C} \alpha_j^{(\ell)} \kappa^{(\ell)} m^{(\ell)} - \kappa^{(j)} m^{(j)} + v_{IN}^{(j)}, \quad j = 0, \ldots, C \qquad (4.21)$$

Let us collect the overall mass and the real inlet mass flow of every balance volume into vectors M and V_{IN} respectively:

$$M = [\, m^{(1)} \ \ldots \ m^{(C)}]^T, \quad V_{IN} = [\, v_{IN}^{(1)} \ \ldots \ v_{IN}^{(C)}]^T$$

Equations (4.21) then can be written in the following matrix-vector form:

$$\frac{dM}{dt} = C_{conv} \mathcal{K} M + V_{IN} \qquad (4.22)$$

where $\mathcal{K} = diag[\kappa^{(j)} \mid j = 1, \ldots, C]$ is a diagonal matrix with positive elements.

Passive Mass Convection Network. We can generalize the above free convection case to obtain a convection network which is asymptotically stable, that is, passive in itself, as follows. Let us collect the mass in- and outflows of every balance volume into the vectors:

$$V_{out} = [\, v_{out}^{(1)} \, \cdots \, v_{out}^{(C)}]^T, \quad V_{in} = [\, v_{in}^{(1)} \, \cdots \, v_{in}^{(C)}]^T$$

and write Equation (4.17) as

$$V_{in} = (C_{conv} + I)V_{out} + V_{IN} \tag{4.23}$$

where I is the unit matrix. In order to make the overall mass subsystem stable, let us apply a full state feedback stabilizing controller (see Chapter 9) in the form of

$$V_{out} = \mathcal{K}M \tag{4.24}$$

with a positive definite square state feedback matrix $\mathcal{K}$. By substituting Equations (4.23) and (4.24) into the overall mass balance equations in Equations (4.21), a linear time-invariant state equation results, which has exactly the same form as Equation (4.22).

Definition 4.2.2 (Passive mass convection network)
An overall mass subsystem with the linear time-invariant state equation (4.22) and with a positive definite square state feedback matrix $\mathcal{K}$ is called a passive mass convection network.

4.3 Special Nonlinear Process Systems

The previous section shows that the state-space model of lumped process systems indeed exhibits a number of special and potentially useful properties. This section is devoted to two interesting special cases within the class of lumped process systems, which are as follows:

- lumped process systems with no source terms, resulting in state equations in bilinear form,
- lumped process systems not obeying the "general modeling assumptions", giving rise to non-standard DAE system models.

4.3.1 Bilinear Process Systems

We have seen in the previous section that lumped process models obeying the "general modeling assumptions" can be transformed into an input-affine nonlinear state-space model form. Moreover, a decomposition of the state equation of lumped process systems has been introduced in Section 4.2.3 in the form of Equation (4.15). There, it was assumed that the flow rates $v^{(j)}$

(and the conserved extensive quantities at the inlet) were the input variables (manipulable input variables or disturbances).

This decomposition shows that the input function g in the input-affine state equation $\dot{x} = f(x) + g(x)u$ is always a *linear* function of the state vector x because of the properties of the convective terms from which it originates. The nonlinear state function $f(x)$ is broken down into a linear term originating from transfer and a general nonlinear term caused by the sources (other generation and consumption terms, including chemical reactions, phase changes, *etc.*). These observations together result in the conditions for a lumped process system model to be in a bilinear form.

Lumped process models with no source term and obeying the "general modeling assumptions" can be transformed into a bilinear state-space model form, assuming that the flow rates (and possibly the conserved extensive quantities at the inlet) are the input variables (manipulable input variables or disturbances).

4.3.2 Process Models in DAE Form

In this subsection we temporarily relax the constant pressure and constant physico-chemical properties assumptions in the set of "general modeling assumptions" in Section 4.1.1 in order to investigate their effect on the set of model equations. It is shown below in the example of a simple evaporator that the presence of the state equation (think of the ideal gas equation as an example), together with the dependence of specific heat capacity on the temperature and pressure, results in a model where the algebraic constitutive equations cannot be substituted into the differential ones. This means that one cannot transform the lumped process model into a canonical nonlinear state-space model form. It remains inherently in its original DAE form.

Example 4.3.1 (A simple evaporator model)

Consider a simple single component phase equilibrium system where vapor and liquid phases are present [31]. This is shown in Figure 4.1. Vapor (denoted by subscript V) and liquid (L) are taken from the vessel, whilst energy is supplied via a heater. Inside the vessel we have two phases with respective hold-ups M_V, M_L and temperatures T_V, T_L. A feed (with mass flow rate F) enters the system. In this model representation, we consider two distinct balance volumes: one for vapor, the other for liquid. From this description we can write the system equations in lumped parameter form, which describes the dynamic behavior:

---

1. Conservation balances
Mass:

$$\frac{dM_V}{dt} = E - V \tag{4.25}$$

$$\frac{dM_L}{dt} = F - E - L \tag{4.26}$$

Energy:

$$\frac{dU_V}{dt} = Eh_{LV} - Vh_V + Q_E \tag{4.27}$$

$$\frac{dU_L}{dt} = Fh_F - Eh_{LV} - Lh_L + Q - Q_E \tag{4.28}$$

2. Transfer rate equations
Mass:

$$E = (k_{LV} + k_{VL})A(P^* - P) \tag{4.29}$$

Energy:

$$Q_E = (u_{LV} + u_{VL})A(T_L - T_V) \tag{4.30}$$

where the subscript LV means liquid to vapor and VL stands for vapor to liquid in the mass and heat transfer coefficients k_i and u_i. The coefficients for VL and LV are normally different.

3. Property relations

$$h_V = \mathbf{h_V}(T_V, P) \tag{4.31}$$

$$h_L = \mathbf{h_L}(T_L, P) \tag{4.32}$$

$$h_{LV} = \mathbf{h_{LV}}(T_L, P) \tag{4.33}$$

$$h_F = \mathbf{h_F}(T_F, P) \tag{4.34}$$

$$P^* = \mathbf{P^*}(T_L) \tag{4.35}$$

$$PV_V = \frac{M_V}{m_w} RT_V \tag{4.36}$$

$$U_V = M_V \mathbf{h_V}(T_V, P) \tag{4.37}$$

$$U_L = M_L \mathbf{h_L}(T_L, P) \tag{4.38}$$

$$V_L = \frac{M_L}{\rho_L} \qquad (4.39)$$

$$k_{LV} = \mathbf{k_{LV}}(T_L, T_V, P) \qquad (4.40)$$

$$k_{VL} = \mathbf{k_{VL}}(T_L, T_V, P) \qquad (4.41)$$

$$u_{LV} = \mathbf{u_{LV}}(T_L, T_V, P) \qquad (4.42)$$

$$u_{VL} = \mathbf{u_{VL}}(T_L, T_V, P) \qquad (4.43)$$

$$\rho_L = \boldsymbol{\rho_L}(T_L, P) \qquad (4.44)$$

4. Balance volume relations

$$V_V = V_T - V_L \qquad (4.45)$$

where V_T is the total volume occupied by the vapor and liquid balance volumes. (In this case, it is the vessel volume.)

5. Equipment and control relations

$$L = f_1(M_L, P) \quad \text{or} \quad L = f_2(M_L) \qquad (4.46)$$

6. Notation
Boldface variables denote thermodynamic property functions. Other system variables are:

M_V	mass hold-up of vapor	M_L	mass holdup of liquid
U_V	vapor phase internal energy	U_L	liquid phase internal energy
F	feed flow rate	V	vapor flow rate
L	liquid flow rate	E	inter-phase mass flow rate
T_V	vapor phase temperature	T_L	liquid phase temperature
Q	energy input flow rate	Q_E	inter-phase energy flowrate
P	system pressure	P^*	vapor pressure
A	interfacial area	R	gas constant
V_V	vapor phase volume	V_T	vessel volume
h_V	vapor specific enthalpy	h_L	liquid specific enthalpy
h_F	feed specific enthalpy	m_w	molecular weight
ρ_L	liquid density	h_{LV}	inter-phase vapor specific enthalpy
V_L	liquid phase volume		

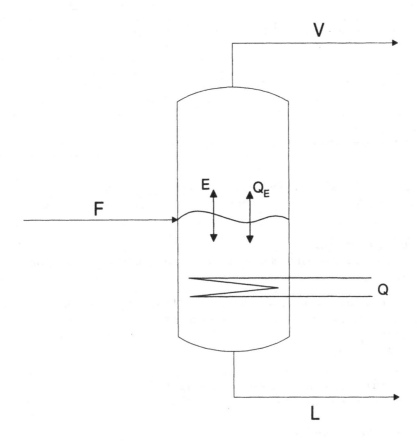

Figure 4.1. A simple evaporator

4.4 Heat Exchanger Examples

Heat exchangers are one of the simplest units in process industries – they can be found in almost every plant. As their name suggests, heat exchangers are used for energy exchange between at least two fluid phase (gas or liquid) streams, a hot and a cold stream. Heat exchangers are usually distributed parameter process systems, but we can build approximate lumped parameter models of them using finite difference approximations of their spatial variables (as in the method of lines approximation scheme). A heat exchanger can then be seen as a composite lumped parameter process system consisting of elementary dynamic units as is depicted in Figure 4.2.

4.4.1 Heat Exchanger Cells

A heat exchanger cell is a primitive dynamic unit which consists of two perfectly stirred (lumped) balance volumes (called *lumps*) connected by a heat

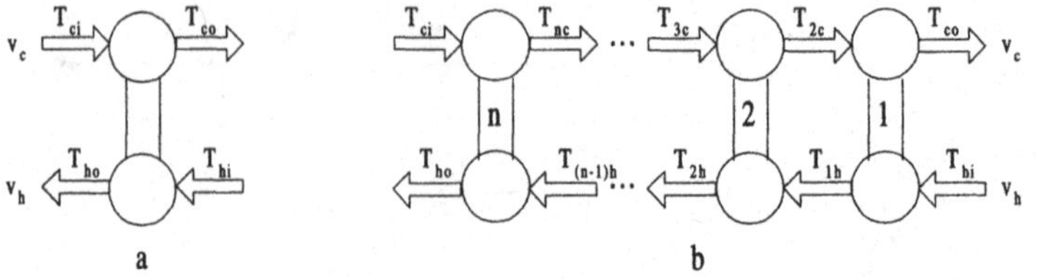

Figure 4.2. A cascade model of a heat exchanger

conducting wall. We shall call one of the lumps the hot ($j = h$) side and the other one the cold ($j = c$) side. The lumps with their variables are shown in Figure 4.2.(/a)

1. Modeling assumptions
In order to obtain a simple model with only two state equations, the following simplifying modeling assumptions are used:

1. Constant volume and mass hold-up in both of the lumps $j = c, h$.
2. Constant physico-chemical properties, such as
   - density: ρ_j
   - specific heat: c_{Pj}
   for both lumps, i.e. for $j = c, h$.
3. Constant heat transfer coefficient (U) and area (A).
4. Completely observable states, i.e. $y(t) = x(t)$.

2. Conservation balances
The continuous time state equations of the heat exchanger cell above are the following energy conservation balances:

$$\dot{T}_{co}(t) = \frac{v_c(t)}{V_c}(T_{ci}(t) - T_{co}(t)) + \frac{UA}{c_{pc}\rho_c V_c}(T_{ho}(t) - T_{co}(t)) \tag{4.47}$$

$$\dot{T}_{ho}(t) = \frac{v_h(t)}{V_h}(T_{hi}(t) - T_{ho}(t)) + \frac{UA}{c_{ph}\rho_h V_h}(T_{co}(t) - T_{ho}(t)) \tag{4.48}$$

where T_{ji} and T_{jo} are the inlet and outlet temperatures, V_j is the volume and v_j is the volumetric flow rate of the two sides ($j = c, h$) respectively.

3. System variables
The *state vector* is therefore composed of the two outlet temperatures:

$$x_1 := T_{co}, \quad x_2 := T_{ho} \tag{4.49}$$

There are a number of possibly time-dependent variables on the right-hand side of the above equations which may act as manipulable *input variables* or disturbances, depending on the measurement and actuator settings and on any additional modeling assumptions we may have. These are as follows:

- the inlet temperatures: T_{ci} and T_{hi},
- the volumetric flowrates: v_c and v_h.

The special cases of the heat exchanger cell models are obtained by specifying assumptions on their variation in time. For every case, the output equation is

$$y(t) = h(x(t)) = \begin{bmatrix} x_1(t) \\ x_2(t) \end{bmatrix} \tag{4.50}$$

4.4.2 LTI State-space Model of a Heat Exchanger Cell

4. Additional modeling assumptions
In order to obtain a finite dimensional linear time-invariant model in each of the cases, the following additional assumptions are applied:

5. Constant volumetric flow rates.
6. Manipulable inlet temperatures.

5. State equations
With assumptions 5 and 6 above, Equations (4.47)–(4.48) become the following finite dimensional LTI state equations:

$$\dot{x} = \begin{bmatrix} -\dfrac{v_c}{V_c} - \dfrac{UA}{c_{Pc}\rho_c V_c} & \dfrac{UA}{c_{Pc}\rho_c V_c} \\ \dfrac{UA}{c_{Ph}\rho_h V_h} & -\dfrac{v_h}{V_h} - \dfrac{UA}{c_{Ph}\rho_h V_h} \end{bmatrix} x + \begin{bmatrix} \dfrac{v_c}{V_c} & 0 \\ 0 & \dfrac{v_h}{V_h} \end{bmatrix} u \tag{4.51}$$

with

$$u = \begin{bmatrix} T_{ci} \\ T_{hi} \end{bmatrix} \tag{4.52}$$

We may divide the state and input matrices in the above equations into additive terms related to the underlying mechanisms as follows. The state matrix term originating from energy transfer is

$$A^{(tr)} = \begin{bmatrix} -\dfrac{UA}{c_{Pc}\rho_c V_c} & \dfrac{UA}{c_{Pc}\rho_c V_c} \\ \dfrac{UA}{c_{Ph}\rho_h V_h} & -\dfrac{UA}{c_{Ph}\rho_h V_h} \end{bmatrix} \tag{4.53}$$

The input and output convection to the lumps gives rise to the following terms in the state and input matrices respectively:

$$A^{(oconv)} = B^{(iconv)} = \begin{bmatrix} \dfrac{v_c}{V_c} & 0 \\ 0 & \dfrac{v_h}{V_h} \end{bmatrix} \tag{4.54}$$

Using these matrices, we can write

$$A = A^{(tr)} - A^{(oconv)}, \quad B = B^{(iconv)} \tag{4.55}$$

4.4.3 LTV State-space Model of a Heat Exchanger Cell

4. Additional modeling assumptions
Let us change the additional assumptions of the previous Subsection 4.4.2 to have:

5. Time-varying volumetric flow rates as disturbances.
6. Manipulable inlet temperatures.

5. State equations
If one cannot assume constant volumetric flow rate, and only the inlet temperatures are manipulable, then the flow rates $v_c(t)$ and $v_h(t)$ can be regarded as time-varying parameters in the LTI model equations in Equation (4.51). The transfer state matrix $A^{(tr)}$ in Equation (4.53) then remains unchanged, but the convection state matrix and the input matrix change to

$$A^{(oconv)}(t) = B^{(iconv)}(t) = \begin{bmatrix} \frac{v_c(t)}{V_c} & 0 \\ 0 & \frac{v_h(t)}{V_h} \end{bmatrix} \qquad (4.56)$$

The decomposition of the state-space model matrices does not change, but we have to use the new convection matrices to get

$$A = A^{(tr)} - A^{(oconv)}(t), \quad B = B^{(iconv)}(t) \qquad (4.57)$$

Note that the state and input vectors are exactly the same as they were in the case of the LTI model.

4.4.4 Nonlinear State-space Model of a Heat Exchanger Cell

In realistic cases, however, the assumptions leading to linear models are not valid, and the flow rates are the manipulable input variables.

4. Additional modeling assumptions
In order to obtain the simplest possible model from this class, the following additional assumptions are used:

5. Manipulable volumetric flow rates,
6. Constant input temperatures.

5. State equations
For the sake of simplicity let us introduce the following notation:

$$u_1 := v_c, \quad u_2 := v_h \qquad (4.58)$$

$$k_1 := \frac{UA}{c_{pc}\rho_c V_c}, \quad k_2 := \frac{UA}{c_{ph}\rho_h V_h} \qquad (4.59)$$

Equations (4.47)–(4.48) can now be written as

$$\dot{x}_1(t) = -k_1 x_1(t) + k_1 x_2(t) + \left(\frac{T_{ci}}{V_c} - \frac{1}{V_c} x_1(t)\right) u_1(t) \tag{4.60}$$

$$\dot{x}_2(t) = k_2 x_1(t) - k_2 x_2(t) + \left(\frac{T_{hi}}{V_h} - \frac{1}{V_h} x_2(t)\right) u_2(t) \tag{4.61}$$

The functions $g_0(x), g_1(x), g_2(x)$ are written (with the t-arguments suppressed) as

$$g_0(x) = f(x) = \begin{bmatrix} -k_1 x_1 + k_1 x_2 \\ k_2 x_1 - k_2 x_2 \end{bmatrix} \tag{4.62}$$

$$g_1(x) = \begin{bmatrix} \frac{T_{ci}}{V_c} - \frac{1}{V_c} x_1 \\ 0 \end{bmatrix} \tag{4.63}$$

$$g_2(x) = \begin{bmatrix} 0 \\ \frac{T_{hi}}{V_h} - \frac{1}{V_h} x_2 \end{bmatrix} \tag{4.64}$$

With the notation above we may write the state equations (4.60) and (4.61) in the following more general form:

$$\dot{x} = A^{(tr)}x + \sum_{i=1}^{2} N_i x u_i + Bu \tag{4.65}$$

6. Mechanisms in the state equations
In the above equation one can clearly see the origin of the terms on the right-hand side:

- linear state *transfer term* $A^{(tr)}$ where

$$A^{(tr)} = \begin{bmatrix} -\frac{UA}{c_{Pc}\rho_c V_c} & \frac{UA}{c_{Pc}\rho_c V_c} \\ \frac{UA}{c_{Ph}\rho_h V_h} & -\frac{UA}{c_{Ph}\rho_h V_h} \end{bmatrix} = \begin{bmatrix} -k_1 & k_1 \\ k_2 & -k_2 \end{bmatrix} \tag{4.66}$$

is a negative semi-definite singular matrix with one zero eigenvalue,
- bilinear state *convection term* originating from the output convection with

$$N_1 = \begin{bmatrix} -\frac{1}{V_c} & 0 \\ 0 & 0 \end{bmatrix}, \quad N_2 = \begin{bmatrix} 0 & 0 \\ 0 & -\frac{1}{V_h} \end{bmatrix}, \tag{4.67}$$

- linear *input term* originating from the input convection with

$$B = \begin{bmatrix} \frac{T_{ci}}{V_c} & 0 \\ 0 & \frac{T_{hi}}{V_h} \end{bmatrix} \tag{4.68}$$

7. Extensions
It is important to note that the form in Equation (4.65) of the bilinear state-space model remains essentially the same if one considers the inlet temperatures to be time-varying.

- If their time variation is modeled via time-varying parameters, then the B matrix above becomes time-varying.
- If they are regarded as additional manipulable input variables, then the input vector u is extended with two new suitably transformed input variables:

$$u = \begin{bmatrix} v_c \\ v_h \\ T_{ci}v_c \\ T_{hi}v_h \end{bmatrix} \tag{4.69}$$

4.5 CSTR Examples

Continuously stirred tank reactors (CSTRs) are the simplest operating units in process systems engineering, because they consist of only a single perfectly stirred balance volume. Therefore the engineering model of a CSTR contains precisely $(K + 1)$ differential equations, being the lumped conservation balances for the overall mass, energy and the K components respectively, supplemented by suitable algebraic constitutive equations.

The state equations are then derived from dynamic conservation balances of the overall mass and component masses, and energy if applicable, by substituting the algebraic equations into the differential ones, if it is possible.

This section contains two CSTR examples. The first one is a simple unstable process system with an autocatalytic chemical reaction. The other one is a simple fermenter model, which we consider in both fed-batch and continuous operation mode.

4.5.1 A Simple Unstable CSTR Example

Let us consider an isothermal CSTR with fixed mass hold-up m and constant physico-chemical properties. A second-order autocatalytic reaction

$$2A + S \rightarrow T + 3A$$

takes place in the reactor, where the substrate S is present in great excess. Assume that the inlet concentration of component A (c_{Ain}) is constant and the inlet mass flow rate v is used as the input variable.

1. Conservation balance equation
The state equation is a single component mass conservation balance equation for component A in the form:

$$\frac{dm_A}{dt} = \frac{d(m \cdot c_A)}{dt} = vc_{Ain} - vc_A + k \cdot m \cdot c_A^2 \tag{4.70}$$

where k is the reaction rate constant. Note that we only have a single balance volume, therefore $C = 1$.

In order to obtain a nonlinear state equation in its input-affine form, we need to *center* the conservation balance equation above. For this purpose a nominal steady-state is used as a reference state. The steady-state values of the system variables are denoted by a superscript asterisk. A given steady-state concentration c_A^* with a nominal mass flow rate v^* satisfies:

$$0 = v^*(c_{Ain} - c_A^*) + k \cdot m \left(c_A^*\right)^2$$

From this we can determine v^* as

$$v^* = -\frac{k \cdot m \left(c_A^*\right)^2}{c_{Ain} - c_A^*}$$

which should be non-negative, therefore $c_{Ain} \leq c_A^*$ should hold. The given steady-state concentration c_A^* also determines the nominal value of the conserved extensive quantity m_A, being the component mass in this case:

$$m_A^* = m \cdot c_A^*$$

4.5.2 A Simple Fed-batch Fermenter

Fermenters are special bio-reactors of great practical importance. They are mostly regarded as CSTRs in which special fermentation reactions take place. Fermentation reaction rate expressions are usually quite complex and highly nonlinear. They are often the main source of nonlinearity in dynamic fermenter models.

1. Modeling assumptions
The simplest dynamic model of a fed-batch fermenter consists of three conservation balances for the mass of the cells (*e.g.* yeast to be produced), that of the substrate (*e.g.* sugar which is consumed by the cells) and for the overall mass. Here we assume that the fermenter is operating under isothermal conditions, so that no energy balance is needed. The cell growth rate is described by a nonlinear static function μ. The particular nature of a fermentation model appears in this so-called source function, which is highly nonlinear and non-monotonous in nature.

Figure 4.3 shows the shape of the nonlinear reaction rate function. The maximum type character of the curve, that is, the non-monotonous property, is the one which makes the system difficult to control.

Initially a solution containing both substrate and cells is present in the fermenter. During the operation we feed a solution of substrate with a given feed concentration S_f to the reactor.

2. State equations
Under the above assumptions the nonlinear state-space model of the fermentation process can be written in the following input-affine form [44]:

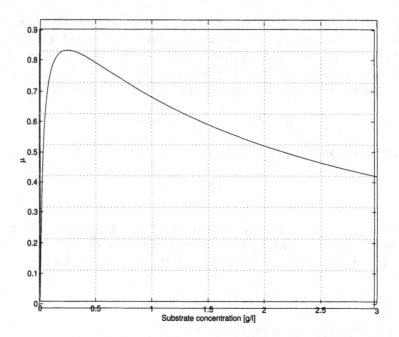

Figure 4.3. The μ function

$$\dot{x} = f(x) + g(x)u \qquad (4.71)$$

where

$$x = \begin{bmatrix} x_1 \\ x_2 \\ x_3 \end{bmatrix} = \begin{bmatrix} X \\ S \\ V \end{bmatrix}, \quad u = F \qquad (4.72)$$

$$f(x) = \begin{bmatrix} \mu(x_2)x_1 \\ -\frac{1}{Y}\mu(x_2)x_1 \\ 0 \end{bmatrix} = \begin{bmatrix} \frac{\mu_{max}x_2x_1}{k_1+x_2+k_2x_2^2} \\ -\frac{\mu_{max}x_2x_1}{(k_1+x_2+k_2x_2^2)Y} \\ 0 \end{bmatrix}, \quad g(x) = \begin{bmatrix} -\frac{x_1}{x_3} \\ \frac{S_f-x_2}{x_3} \\ 1 \end{bmatrix} \qquad (4.73)$$

and

$$\mu(x_2) = \frac{\mu_{max}x_2}{k_1 + x_2 + k_2x_2^2} \qquad (4.74)$$

3. System variables
The variables of the model and their units are the following:

$x_1 = X$	cell concentration (state)	[g/l]
$x_2 = S$	substrate concentration (state)	[g/l]
$x_3 = V$	volume (state)	[l]
$u = F$	feed flow rate (input)	[l/h].

4. Model parameters

A typical set of constant parameters and their values is as follows:

$Y = 0.5$ yield coefficient

$\mu_{max} = 1$ maximum growth rate $[h^{-1}]$

$k_1 = 0.03$ Monod constant $[g/l]$

$k_2 = 0.5$ kinetic parameter $[l/g]$

$S_f = 10$ influent substrate concentration $[g/l]$

$X_f = 0$ influent cell concentration $[g/l]$

c_1, c_2 reaction enthalpy coefficients.

4.5.3 Simple Continuous Fermenter Models

For convenience, a relatively simple bio-reactor that is similar to the fed-batch case is selected but is now operated in continuous mode.

Despite its simplicity, this model exhibits some of the key properties which render bio-reactors difficult to operate, and therefore proper controller design for bio-reactors is important.

1. Modeling assumptions

The continuous reactor is assumed to be perfectly stirred and an unstructured biomass growth rate model with substrate inhibition kinetics is chosen.

The fermenter is considered to be isothermal with constant volume V and constant physico-chemical properties.

2. Conservation balances

The dynamics of the process are given by the state-space model derived from the conservation balances for the biomass with concentration X and substrate mass with concentration S.

$$\frac{dX}{dt} = \mu(S)X - \frac{XF}{V} \tag{4.75}$$

$$\frac{dS}{dt} = -\frac{\mu(S)X}{Y} + \frac{(S_F - S)F}{V} \tag{4.76}$$

$$\text{where } \mu(S) = \mu_{max}\frac{S}{K_2 S^2 + S + K_1} \tag{4.77}$$

The first equation originates from the biomass component mass balance, while the second is from the substrate component mass balance. They are coupled by the nonlinear growth rate function $\mu(S)X$, which is the main source of nonlinearity and uncertainty in this simple model.

3. System variables and model parameters

The variables and parameters of the model together with their units and parameter values are given in Table 4.1. The parameter values are taken from [44].

4. State equations

The above model can easily be written in standard input-affine form with the centered state vector

Table 4.1. Variables and parameters of the fermentation process model

X	biomass concentration		[g/l]
S	substrate concentration		[g/l]
F	feed flow rate		[l/h]
V	volume	4	[l]
S_F	substrate feed concentration	10	[g/l]
Y	yield coefficient	0.5	-
μ_{max},	maximal growth rate	1	[1/h]
K_1	saturation parameter	0.03	[g/l]
K_2	inhibition parameter	0.5	[l/g]

$$x = [\bar{X} \ \bar{S}]^T = [X - X_0 \ S - S_0]^T$$

consisting of the centered biomass and substrate concentrations. The centered input flow rate is chosen as manipulable input variable, *i.e.*

$$u = \bar{F} = F - F_0$$

Then we have

$$\dot{x} = f(x) + g(x)u \tag{4.78}$$

$$f(x) = \begin{bmatrix} \mu(\bar{S} + S_0)(\bar{X} + X_0) - \frac{(\bar{X}+X_0)F_0}{V} \\ -\frac{\mu(\bar{S}+S_0)(\bar{X}+X_0)}{Y} + \frac{(S_F-(\bar{S}+S_0))F_0}{V} \end{bmatrix}$$

$$g(x) = \begin{bmatrix} -\frac{(\bar{X}+X_0)}{V} \\ \frac{(S_F-(\bar{S}+S_0))}{V} \end{bmatrix} \tag{4.79}$$

with (X_0, S_0, F_0) being a steady-state operating point.

5. Calculation of the optimal operating point

The maximal biomass productivity XF is selected as the desired optimal operating point, *i.e.* the substrate cost is assumed to be negligible. This equilibrium point can be calculated from the nonlinear model:

$$S_0 = \frac{1}{2} \frac{-2K_1 + 2\sqrt{K_1^2 + S_F^2 K_1 K_2 + S_F K_1}}{S_F K_2 + 1} \tag{4.80}$$

$$X_0 = (S_F - S_0)Y \tag{4.81}$$

and the corresponding inlet feed flow rate is

$$F_0 = \mu(S_0)V \tag{4.82}$$

Substituting the parameter values from Table 4.1 gives

$$S_0 = 0.2187 \text{ g/l}, \ X_0 = 4.8907 \text{ g/l}, \ F_0 = 3.2029 \text{ l/h} \tag{4.83}$$

6. Linearized model

In order to compare linear and nonlinear model analysis and control techniques, the linearized version of the nonlinear model – state equations are in Equations (4.78)–(4.79) – is presented here:

$$\dot{x} = Ax + Bu \tag{4.84}$$

where

$$A = \left[\frac{\partial f}{\partial x}\right]_{x=0} = \begin{bmatrix} 0 & -\frac{\mu_{max}X_0(K_2S_0^2-K_1)}{(K_2S_0^2+S_0+K_1)^2} \\ -\frac{\mu_{max}S_0}{(K_2S_0^2+S_0+K_1)Y} & \frac{\mu_{max}X_0(K_2S_0^2-K_1)}{(K_2S_0^2+S_0+K_1)^2} - \frac{F_0}{V} \end{bmatrix} \tag{4.85}$$

$$B = g(0) = \begin{bmatrix} -\frac{X_0}{V} \\ \frac{S_F-S_0}{V} \end{bmatrix} \tag{4.86}$$

The steady-state point (X_0, S_0, F_0) has been used in the linearized version of the model. The system matrices at the optimal operating point are:

$$A = \begin{bmatrix} 0 & 0.4011 \\ -1.6045 & 1.2033 \end{bmatrix}, \quad B = \begin{bmatrix} -1.227 \\ 2.4453 \end{bmatrix} \tag{4.87}$$

4.6 Case Study: Modeling a Gas Turbine

Gas turbines are important and widely used prime movers in transportation systems such as aircraft and cars. They are also found in power systems, where they are the main power generators, and in process plants as well. The investigation of steady-state behavior and static characteristics of gas turbines is a traditional area in engineering. This kind of model is based upon the characteristics of the component parts of the engine. The static characteristics can be given in the form of polynomials reflecting the results of the preliminary calculations or the measurements.

This section is devoted to the model building of a low-power gas turbine [1], which will be used later on in the book as a case study for nonlinear analysis and control.

4.6.1 System Description

The main parts of a gas turbine include the compressor, the combustion chamber and the turbine. For a jet engine, there is also the inlet duct and the nozzle. The interactions between these components are fixed by the physical structure of the engine. The operation of these two types of gas turbine is basically the same. The air is drawn into the engine by the compressor, which compresses it and then delivers it to the combustion chamber. Within the combustion chamber the air is mixed with fuel and the mixture is ignited, producing a rise in temperature and hence an expansion of the gases. These gases are exhausted through the engine nozzle or the engine gas-deflector, but first pass through the turbine, which is designed to extract sufficient energy from them to keep the compressor rotating so that the engine is self-sustaining. The main parts of the gas turbine are shown schematically in Figure 4.4. In this section we analyze a low-power gas turbine, which is in-

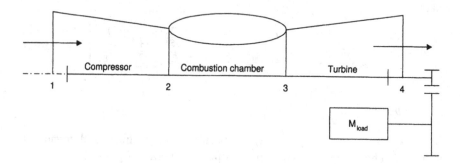

Figure 4.4. The main parts of a gas turbine

stalled on a test-stand in the Technical University of Budapest, Department of Aircraft and Ships.

Engineering intuition suggests that the gas turbine is inherently stable because of its special dynamics. This is also confirmed by our measurements and simulation results. The most important control aim could then be to keep the number of revolutions constant, unaffected by the load and the ambient conditions (pressure and temperature). The temperatures and the number of revolution has to be limited, their values are bounded from above by their maximum values.

4.6.2 Modeling Assumptions

In order to get a low order dynamic model suitable for control purposes, some simplifying modeling assumptions should be made.

General assumptions that apply in every section of the gas turbine

1. Constant physico-chemical properties are assumed. These include the specific heats at constant pressure and at constant volume, the specific gas constant and adiabatic exponent.
2. Perfectly stirred balance volumes (lumps) are assumed in each main part of the gas turbine. This means that a finite dimensional concentrated parameter model is developed and the values of the variables within a balance volume are equal to those at the outlet.

Other assumptions

3. Efficiency of the combustion is constant.
4. In the compressor and in the turbine the mass flow rates are constant:

$$\nu_{Cin} = \nu_{Cout} = \nu_C \quad \text{and} \quad \nu_{Tin} = \nu_{Tout} = \nu_T$$

4.6.3 Conservation Balances

The nonlinear state equations are derived from first engineering principles. Dynamic conservation balance equations are constructed for the overall mass m and internal energy U for each of the three main parts of the turbine system [32]. The notation list is given separately in Table 4.2.

Table 4.2. The variables and parameters of the gas turbine model

Variables		Indices	
m	mass	$comb$	combustion chamber
U	internal energy	$fuel$	refers to the fuel
T	temperature	C	compressor
p	pressure	T	turbine
n	number of revolution	in	inlet
c	specific heat	out	outlet
i	enthalpy	p	refers to constant pressure
M	moment	v	refers to constant volume
R	specific gas constant	$comb$	refers to combustion
η	efficiency	$mech$	mechanical
Θ	inertial moment	$load$	loading
ν	mass flow rate	air	refers to air
		gas	refers to gas

The development of the model equations is performed in the following steps:

1. *Conservation balance for total mass (applies to each section of the gas turbine):*

$$\frac{dm}{dt} = \nu_{in} - \nu_{out} \tag{4.88}$$

2. *Conservation balance for total energy in each section of the gas turbine,* where the heat energy flows and work terms are also taken into account:

$$\frac{dU}{dt} = \nu_{in} i_{in} - \nu_{out} i_{out} + Q + W \tag{4.89}$$

We can transform the above energy conservation equation by considering the dependence of the internal energy on the measurable temperature:

$$\frac{dU}{dt} = c_v \frac{d}{dt}(Tm) = c_v T \frac{dm}{dt} + c_v m \frac{dT}{dt} \tag{4.90}$$

From the two equations above we get a state equation for the temperature:

$$\frac{dT}{dt} = \frac{v_{in} i_{in} - v_{out} i_{out} + Q + W - c_v T (v_{in} - v_{out})}{c_v m} \tag{4.91}$$

3. The *ideal gas equation* is used as a constitutive equation together with two balance equations above to develop an alternative state equation for the pressure:

$$\frac{dp}{dt} = \frac{RT}{V}(v_{in} - v_{out})$$

$$+ \frac{p}{T}\left(\frac{v_{in} i_{in} - v_{out} i_{out} + Q + W - c_v T (v_{in} - v_{out})}{c_v m}\right) \tag{4.92}$$

Note that both the extensive and the intensive forms of the model equations are used later for model analysis.

4.6.4 Conservation Balances in Extensive Variable Form

The state equations in extensive variable form include the dynamic mass conservation balance for the combustion chamber, the internal energy balances for all of the three main parts of the turbine and an overall energy balance for the system originating from the mechanical part. The indices in the balance equations and variables therein refer to the main parts of the turbine: to the compressor ($i = 2$), to the combustion chamber ($i = 3$) and to the turbine ($i = 4$), while the inlet variables are indexed by $i = 1$.

Thus five independent balance equations can be constructed, therefore the gas turbine can be described by only five state variables.

Total mass balance

$$\frac{dm_{Comb}}{dt} = v_C + v_{fuel} - v_T \tag{4.93}$$

Total energy balance

$$\frac{dU_2}{dt} = v_C c_{pair}(T_1 - T_2) + v_T c_{pgas}(T_3 - T_4)\eta_{mech} - 2\Pi \frac{3}{50} n M_{load} \tag{4.94}$$

$$\frac{dU_3}{dt} = v_C c_{pair} T_2 - v_T c_{pgas} T_3 + Q_f \eta_{comb} v_{fuel} \tag{4.95}$$

$$\frac{dU_4}{dt} = v_T c_{pgas}(T_3 - T_4) - \frac{v_C c_{pair}(T_2 - T_1)}{\eta_{mech}} - \frac{2\Pi}{\eta_{mech}} \frac{3}{50} n M_{load} \tag{4.96}$$

Mechanical dynamic equation

$$\frac{dP_n}{dt} = v_T c_{pgas}(T_3 - T_4)\eta_{mech} - v_C c_{pair}(T_2 - T_1) - 2\Pi \frac{3}{50} n M_{load} \tag{4.97}$$

4.6.5 Model Equations in Intensive Variable Form

There are several alternatives for the model equations in intensive variable form. We choose the set that includes the dynamic mass balance for the combustion chamber (that is, Equation (4.93)), the pressure form of the state equations derived from the energy balances and the intensive form of the overall mechanical energy balance, expressed in terms of the number of revolutions n.

$$\frac{dp_2}{dt} = \frac{R_{air}}{V_C c_{vair}}(\nu_C c_{pair}(T_1 - \frac{p_2 V_C}{m_C R_{air}}) + \nu_T c_{pgas}(\frac{p_3 V_{Comb}}{m_{Comb} R_{med}}$$

$$-\frac{p_4 V_T}{m_T R_{gas}})\eta_{mech} - 2\Pi\frac{3}{50}n M_{load}) \tag{4.98}$$

$$\frac{dp_3}{dt} = \frac{p_3}{m_{Comb}}(\nu_C + \nu_{fuel} - \nu_T) +$$

$$+\frac{R_{med}}{V_{Comb} c_{vmed}}(\nu_C c_{pair}\frac{p_2 V_C}{m_C R_{air}} - \nu_T c_{pgas}\frac{p_3 V_{Comb}}{m_{Comb} R_{med}}$$

$$+Q_f \eta_{comb}\nu_{fuel} - c_{vmed}\frac{p_3 V_{Comb}}{m_{Comb} R_{med}}(\nu_C + \nu_{fuel} - \nu_T)) \tag{4.99}$$

$$\frac{dp_4}{dt} = \frac{R_{gas}}{V_T c_{vgas}}(\nu_T c_{pgas}(\frac{p_3 V_{Comb}}{m_{Comb} R_{med}}$$

$$-\frac{p_4 V_T}{m_T R_{gas}}) - \frac{\nu_C c_{pair}(\frac{p_2 V_C}{m_C R_{air}} - T_1)}{\eta_{mech}} - \frac{2\Pi}{\eta_{mech}}\frac{3}{50}n M_{load}) \tag{4.100}$$

$$\frac{dn}{dt} = \frac{1}{4\Pi^2\Theta n}(\nu_T c_{pgas}(\frac{p_3 V_{Comb}}{m_{Comb} R_{med}} - \frac{p_4 V_T}{m_T R_{gas}})\eta_{mech}$$

$$-\nu_C c_{pair}(\frac{p_2 V_C}{m_C R_{air}} - T_1) - 2\Pi\frac{3}{50}n M_{load}) \tag{4.101}$$

4.6.6 Constitutive Equations

Two types of constitutive equations are needed to complete the nonlinear gas turbine model. The first is the ideal gas equation

$$T = \frac{pV}{mR}$$

which has already been used before, and has been substituted into the state equations to get alternative intensive forms.

The second type of constitutive equations describes the mass flow rate in the compressor and in the turbine.

$$\nu_C = const(1)q(\lambda_1)\frac{p_1}{\sqrt{T_1}} \tag{4.102}$$

$$\nu_T = \text{const}(2)q(\lambda_3)\frac{p_3}{\sqrt{T_3}} \tag{4.103}$$

In these equations $q(\lambda_1)$ and $q(\lambda_3)$ can be calculated as follows:

$$q(\lambda_1) = f(\frac{n}{\sqrt{T_1}}, \frac{p_2}{p_1}) \tag{4.104}$$

$$q(\lambda_3) = f(\text{const}(3)\frac{n}{\sqrt{\frac{p_3 V_{Comb}}{m_{Comb} R_{med}}}}, \frac{p_3}{p_4}) \tag{4.105}$$

The parameters and constants of these functions can be determined using measured data and the compressor and turbine characteristics.

4.6.7 Operation Domain and System Variables

The *measurable intensive set of state variables* for the gas turbine test stand at the Technical University of Budapest is:

$$\bar{x} = [\, m_{Comb} \quad p_2 \quad p_3 \quad p_4 \quad n \,]^T \tag{4.106}$$

Experimental values of these variables are constrained to the following domain:

$0.003 \le m_{Comb} \le 0.0067$ [kg] $180000 \le p_2^* \le 280000$ [Pa]
$170000 \le p_3^* \le 270000$ [Pa] $100000 \le p_4^* \le 140000$ [Pa]
$39000 \le n \le 51000$ [1/min]

The value of the only *input variable* ν_{fuel} is also constrained by:

$$0.009 \le \nu_{fuel} \le 0.017 \text{ [kg/sec]}$$

The set of possible *disturbances* includes:

$$d = [\, T_1 \quad p_1 \quad M_{terh} \,]^T \tag{4.107}$$

where the domain of its elements is:

$273 \le T_1 \le 310$ [K] $97000 \le p_1 \le 103000$ [Pa] $0 \le M_{terh} \le 180$ [Nm]

Finally we construct the *set of output variables* by noticing that all the pressures p_i^* and the number of revolutions n in the state vector above can be measured, but the mass m_{Comb} cannot:

$$y = [\, p_2 \quad p_3 \quad p_4 \quad n \,] \tag{4.108}$$

4.7 Summary

The basic approach and a systematic methodology for constructing lumped dynamic process models from first engineering principles is introduced in this chapter. The most important mechanisms (convection, transfer and a source term associated with chemical reactions) are also discussed in detail and their effect on the algebraic form of the state equations is also described.

The modeling methodology is illustrated in several case studies of simple, yet practically important, process systems: heat exchangers, continuously stirred tank reactors including bio-reactors, and a gas turbine. The nonlinear state-space models developed here will be used for nonlinear model analysis and control studies throughout the book.

4.8 Questions and Application Exercises

Exercise 4.8.1. Which are the canonical extensive variables and related potentials of lumped process systems?

Exercise 4.8.2. What are the general expression for total mass balance in a process system? Describe the significance of each term.

Exercise 4.8.3. What are the principal mechanisms in lumped process systems? How can we decompose the state equation of a lumped process system driven by mechanisms?

Exercise 4.8.4. What are the most important classes of constitutive equations? Which ones of those play role in dynamic modeling for control?

Exercise 4.8.5. What are the general expressions for energy balance in a process system? Describe the significance of each term.

Exercise 4.8.6. Develop a special case of the simple evaporator model described in Example 4.3.1 assuming constant physico-chemical properties. Construct the state equations of the system by substituting the algebraic constitutive equations into the differential ones.

Exercise 4.8.7. Develop a special case of the simple fed-batch fermenter model described in Subsection 4.5.2 assuming a more simple fermentation kinetics:

$$\mu(S) = \mu_{max} \frac{S}{K_2 S + K_1}$$

Construct the state equations of the system.

Exercise 4.8.8. Construct an LTI state-space model of the heat exchanger cell described in Subsection 4.4.2 using *centered variables*. Compare the obtained model with the one in Equations (4.51).

Exercise 4.8.9. Construct a bilinear state-space model of the heat exchanger cell described in Subsection 4.4.4 using *centered variables*.

Exercise 4.8.10. Develop a linearized state-space model of the nonlinear heat exchanger cell described in Section 4.4.4.

5. Input–output Models and Realization Theory

Basic notions of signals and systems (see Chapter 2) suggest that input–output description is the natural description of dynamic systems represented as an abstract operator **S** acting on the input space and producing a signal from the output space. If we only use the input and output signals (possibly together with disturbances) for the description, then we have an *input–output model* or description of the system in question.

This chapter is devoted to the detailed elaboration of this concept of input–output description both for continuous time finite–dimensional LTI and nonlinear systems.

The material in this chapter is arranged in the following sections:

- *Input–output models of LTI systems*
  The basic notions of both the time domain and frequency (operator) domain description are introduced.
- *Input–output representation of nonlinear systems*
  The basic representation forms of LTI systems are generalized to cover the general finite dimensional nonlinear case. Both the Fliess and Volterra series representations are described.
- *Realization theory*
  A brief introduction is given here on how to find state-space models for a given input–output model in both the LTI and nonlinear case.
- *Zero dynamics*
  The notion of zero dynamics is very useful when designing output feedback controllers to a nonlinear system. It is introduced here and its computation is illustrated for a simple process system example.

5.1 Input–output Models of LTI Systems

Usually we do not use the abstract system operator **S** when we want to describe a concrete system for the purpose of its analysis and control. Instead, we use different other descriptions: *system models* or *system representations* mainly in the form of differential and algebraic equations. The following subsections deal with different input–output and state-space representations of continuous time LTI systems.

For the sake of simplicity we only deal with *single-input– single-output* (SISO) systems here, but the different models can be easily generalized to the MIMO case, as well. A SISO LTI system is usually described either in

- time domain, or in
- operator domain.

5.1.1 Time Domain Description

There are different alternative methods for the input–output description of LTI systems in time domain: linear higher-order differential equations with constant coefficients and the impulse-response function description.

Linear Differential Equations with Constant Coefficients. Let us denote the scalar output signal of the system by y and the scalar input by u. Then the system model is in the form of a linear higher-order ordinary differential equation:

$$a_n \frac{d^n y}{dt^n} + a_{n-1} \frac{d^{n-1} y}{dt^{n-1}} + \ldots + a_1 \frac{dy}{dt} + a_0 y = b_0 u + b_1 \frac{du}{dt} + \ldots + b_m \frac{d^m u}{dt^m} \quad (5.1)$$

with given initial conditions

$$y(0) = y_{00}, \quad \frac{dy}{dt}(0) = y_{10}, \quad \ldots \quad , \quad \frac{d^{n-1} y}{dt^{n-1}}(0) = y_{n0}$$

because the input function u and its derivatives are considered to be known (given) for every time instant. Observe that we need n initial conditions for this description.

The *parameters* of this system model are the constants

$$(a_0, \ a_1, \ \ldots, \ a_n), \ (b_0, \ b_1, \ \ldots, \ b_m)$$

This system model is clearly

- *linear*, because the equation is linear, and
- *time-invariant* because the model parameters are constant (do not depend on time).

Impulse-response Representation.

Definition 5.1.1 (Impulse-response function)
The impulse-response function is the response of a SISO LTI system to a Dirac-δ input function with zero initial condition.

The concept of impulse-response representation is illustrated in Figure 5.1 below where

$h(t)$: impulse-response function
$\delta(t)$: Dirac-δ function.

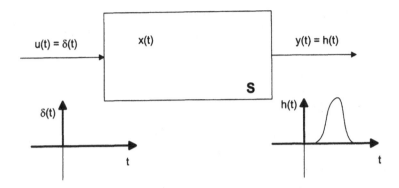

Figure 5.1. Notion of the impulse-response function

The output of **S** can be written as

$$y(t) = \int_{-\infty}^{\infty} h(t - \tau)u(\tau)d\tau = \int_{-\infty}^{\infty} h(\tau)u(t - \tau)d\tau \qquad (5.2)$$

The equation above describes the *convolution* of $h(.)$ and $u(.)$ in time domain.

We can develop two equivalent forms from the above defining equations (5.2) as follows:

1. Because $u(.)$ is a function $\mathbb{R}_0^+ \rightarrow \mathbb{R}$, we start the integration at time 0. The upper bound for the integration is t because the system is *causal*, i.e.

$$y(t) = \int_0^t h(t - \tau)u(\tau)d\tau \qquad (5.3)$$

The above integral formulation is only valid under the so-called "zero initial condition" assumption, when $\{u(t) = 0 \mid t < 0\}$. This condition will be reformulated later on for state-space representations.

2. We can observe that $h(t)$ is also identically zero up to the time $t = 0$ because of the definition and the causality of the system. Therefore Equation (5.2) specializes to

$$y(t) = \int_0^{\infty} h(\tau)u(t - \tau)d\tau \qquad (5.4)$$

5.1.2 Operator Domain Description

In operator domain we work with the Laplace transform (denoted by $\mathcal{L}$, abbreviated by L-transform) of signals. The Laplace transform is defined in Section 2.1.4.

The description of an LTI system in operator domain uses the transfer function $H(s)$ of the system.

Definition 5.1.2 (Transfer function)
The transfer function of a SISO LTI system is a complex function which is defined as follows:

$$Y(s) = H(s)U(s) \tag{5.5}$$

assuming zero initial conditions with

$Y(s)$ *Laplace transform of the output signal*
$U(s)$ *Laplace transform of the input signal*
$H(s) = \frac{b(s)}{a(s)}$ *transfer function of the system*
                 where $a(s)$ and $b(s)$ are polynomials and
                 degree $b(s) = m$
                 degree $a(s) = n$.

We can easily connect the time domain differential equation description (5.1) of a continuous time SISO LTI system with the transfer function description if we take the Laplace transform of Equation (5.1) with zero initial conditions and compare the result with Equation (5.5).

Using property (3) of the Laplace transform and the defining equation of the transfer function (5.2) it is easy to show that $H(s)$ is the Laplace transform of the impulse-response function $h(t)$ of the system **S**:

$$H(s) = \mathcal{L}\{h(t)\} \tag{5.6}$$

Definition 5.1.3 (Proper transfer function)
Let the transfer function of a continuous time SISO LTI system be in the form of $H(s) = \frac{b(s)}{a(s)}$ with $n = \deg(a(s)), m = \deg(b(s))$ being the degrees of the denominator and nominator polynomials respectively. If

$m < n$ *then $H(s)$ is*   strictly proper
$m = n$ proper
$m > n$ improper.

It is important to note that realistic causal systems are strictly proper therefore it will be assumed from now on.

5.1.3 Input–output and State-space Representations of LTI Systems

The concept of state-space models has already been introduced in Chapter 3. It is of great importance to find connections between state-space and input–output realizations of a given LTI system. Here we consider the problem of constructing various forms of input–output models to a given LTI state-space model or realization.

Definition 5.1.4 (Equivalent realizations)
Equivalent state-space realizations (or models) are the ones which give rise to the same input-output description, that is to the same transfer function in the LTI case.

The general form of state-space models of LTI systems has been already introduced in Section 3.2.1 in the form:

$$\dot{x}(t) = Ax(t) + Bu(t), \quad x(0) = x_0$$
$$y(t) = Cx(t) + Du(t)$$

which is characterized by the matrix quadruplet (A, B, C, D) where usually $D = 0$ is assumed.

Construction of the Transfer Function from State-space Models.
Assuming that x(0)=0, we can easily derive the transfer function from the state-space model equation above by computing the Laplace transform of both equations and substituting $X(s)$ from the transformed state equations to the output equation:

$$X(s) = (sI - A)^{-1}BU(s)$$
$$Y(s) = \{C(sI - A)^{-1}B + D\}U(s) \tag{5.7}$$

The last equation gives the transfer function $H(s)$ of the state-space representation matrices (A, B, C, D):

$$H(s) = C(sI - A)^{-1}B + D \tag{5.8}$$

The above derivation shows that the transfer function can be computed from the realization matrices (A, B, C, D) therefore their combination in Equation (5.8) is realization-independent.

There are other more simple quantities which are realization-independent in the case of LTI systems. The most important of them are the Markov parameters defined below.

Definition 5.1.5 (Markov parameters)
The Markov parameters of a continuous time LTI system with state-space realization (A, B, C) are defined as:

$$h_i = CA^{i-1}B, \quad i = 1, 2, ... \tag{5.9}$$

It can be shown that the Markov parameters are invariant under state transformations, that is, they remain unchanged if one applies a state transformation using an invertible transformation matrix T.

Construction of the Impulse-response Function from State-space Models. We can relate the parameters of a given state-space model to the impulse-response function in two ways. The first way is to use the solution of the state equation and remember that the impulse-response function $h(t)$ of a LTI system is its output for the Dirac-δ ($\delta(t)$) input and we obtain with $D = 0$:

$$h(t) = Ce^{At}B = CB + CABt + CA^2B\frac{t^2}{2!} + ... \tag{5.10}$$

The same result is obtained by taking the inverse-Laplace transformation of Equation (5.8) with $D = 0$.

Note that the Markov parameters are present in the time series expansion of the impulse-response function $h(t)$.

5.2 Input–output Representation of Nonlinear Systems

The input–output representation of finite dimensional nonlinear systems is much more difficult than that of LTI systems. Besides the mathematical difficulties present in the various representations, one has to cope with the infinite series inherent in the various representations.

5.2.1 Fliess's Functional Expansion

First, we summarize the necessary mathematical tools and definitions for the input–output description of nonlinear systems based on [37].

Definition 5.2.1 (Multi-index)
Let I_k denote the set of all sequences $(i_k \ldots i_1)$ of k elements $i_k, \ldots, i_1$ of the index set I. A multi-index is an element of I_k. The only element of I_0 is the empty sequence (i.e. a multi-index of length 0), denoted by $\emptyset$.

In the case of an input-affine nonlinear system with m inputs, the index set is $I = \{0, 1, ..., m\}$. Furthermore, let us introduce the following notation:

$$I^* = \bigcup_{k \geq 0} I_k \tag{5.11}$$

Definition 5.2.2 (Formal power series)
A formal power series in $m + 1$ noncommutative indeterminates and coefficients in $\mathbb{R}$ is a mapping

$$c : I^* \to \mathbb{R} \tag{5.12}$$

The value of c at some element $i_k \ldots i_0$ of I^* is denoted by $c(i_k \ldots i_0)$.

Definition 5.2.3 (Iterated integrals)

Let T be a fixed value of the time and suppose $u_1, \ldots, u_m$ are real-valued piece-wise continuous functions defined on $[0, T]$. For each multi-index $(i_k \ldots i_0)$ the corresponding iterated integral is a real-valued function of t

$$E_{i_k \ldots i_1 i_0} = \int_0^t d\xi_{i_k} \ldots d\xi_{i_1} d\xi_{i_0}$$

defined for $0 \leq t \leq T$ by recurrence on the length, setting

$$\xi_0(t) = t$$

$$\xi_i(t) = \int_0^t u_i(\tau) d\tau \quad for \ 1 \leq i \leq m$$

and

$$\int_0^t d\xi_{i_k} \ldots d\xi_{i_0} = \int_0^t d\xi_{i_k}(\tau) \int_0^\tau d\xi_{i_{k-1}} \ldots d\xi_{i_0}$$

The iterated integral corresponding to the multi-index $\emptyset$ is the real number 1.

Example 5.2.1 (The first few iterated integrals)

The iterated integrals are computed in a recursive way following the recursive structure of their definition.

$$\xi_0(t) = t$$

$$\xi_i(t) = \int_0^t u_i(\tau) d\tau \quad for \ 1 \leq i \leq m$$

$$d\xi_i = \frac{d\xi_i}{dt}$$

$$\int_0^t d\xi_0 = t,$$

$$\int_0^t d\xi_1 = \int_0^t u(\tau) d\tau$$

$$\int_0^t d\xi_0 d\xi_0 = \frac{t^2}{2!} \tag{5.13}$$

$$\int_0^t d\xi_0 d\xi_1 = \int_0^t \int_0^\tau u(\theta) d\theta d\tau$$

$$\int_0^t d\xi_1 d\xi_0 = \int_0^t u(\tau) \tau d\tau$$

$$\int_0^t d\xi_1 d\xi_1 = \int_0^t u(\tau) \left[\int_0^\tau u(\theta) d\theta \right] d\tau$$

$$\ldots$$

The Fliess's series expansion makes use of the iterated integrals above to obtain an input–output representation of lumped nonlinear systems in the form

$$y = S[u], \quad u \in \mathcal{U}, \ y \in \mathcal{Y} \tag{5.14}$$

where S is the system operator, u is the input, y is the output, and $\mathcal{U}$ and $\mathcal{Y}$ denote the input and output signal spaces with $\dim(\mathcal{U}) = m$ and $\dim(\mathcal{Y}) = p$ respectively.

Theorem 5.2.1 (Fliess's series expansion of nonlinear systems). *Suppose the inputs $u_1, \ldots, u_m$ of the nonlinear system (5.14) satisfy the constraint*

$$\max_{0 \leq \tau \leq T} |u_i(\tau)| < 1$$

Then, if T is sufficiently small, the j-th output $y_j(t)$ of the system (5.14) may be expanded in the following way:

$$y_j(t) = h_j(x^0) + \sum_{k=0}^{\infty} \sum_{i_0,\ldots,i_k=0}^{m} L_{g_{i_0}} \ldots L_{g_{i_k}} h_j(x^0) \int_0^t d\xi_{i_k} \ldots d\xi_{i_0} \tag{5.15}$$

where $g_0 = f$.

5.2.2 Volterra Series Representation

The convolution integral in (5.2) can be generalized and applied to a large class of nonlinear systems. The result of this generalization is the so-called *Volterra-series* description, where we describe the input–output behavior of the system using a series of so-called generalized convolution integrals. The linear case in Equation (5.4) is a special case of this representation.

For simpler notation, let us consider a nonlinear input-affine system model with scalar output, *i.e.* Equation (3.19) being in the form of

$$\dot{x}(t) = f(x(t)) + \sum_{i=1}^{m} g_i(x(t))u_i(t)$$
$$y(t) = h(x(t))$$

with $\dim y = p = 1$. The output of a system can be approximated with the series

$$y(t) = w_0(t) +$$
$$\sum_{k=1}^{\infty} \int_0^t \int_0^{\tau_1} \ldots \int_0^{\tau_{k-1}} w_k(t, \tau_1, \ldots, \tau_k) u(\tau_1) \ldots u(\tau_k) d\tau_k \ldots d\tau_1 \tag{5.16}$$

where $w_k(t, \tau_1, \ldots, \tau_k)$ is called the *k-th order Volterra kernel*.

Let us denote the flow of the vector field f in Equation (3.19) by f^t (*i.e.* $f^t(x)$ is the solution of the equation above at time t starting from $x(0) = x$

with $u = 0$). It can be shown (see, *e.g.* [21]) that the Volterra kernels are calculated as

$$w_0(t) = h \circ f^t(x_0)$$
$$w_k(t, \tau_1, \ldots, \tau_k) =$$
$$L_g(\ldots(L_g(L_g(h \circ f^{t-\tau_1}) \circ f^{\tau_1-\tau_2}) \circ f^{\tau_2-\tau_3}) \circ \cdots \circ f^{\tau_{k-1}-\tau_k}) \circ f^{\tau_k}(x_0)$$
$$k = 1, 2, \ldots$$

$$(5.17)$$

The main result from the convergence of (5.17) is that if f, g and h are analytic in a neighborhood of x_0 then there exists $T > 0$ such that for each input function satisfying $|u(t)| < 1$ on $[0, T]$, the series (5.17) is uniformly absolutely convergent on $[0, T]$ (see, *e.g.* [21]).

5.2.3 Higher-order Nonlinear Differential Equations

Similarly to the time domain input–output description of LTI systems described in Section 5.1.1, under some conditions the relation between the inputs and outputs of a nonlinear system (3.19) can also be represented using a set of higher-order differential equations of the form

$$F_i\left(u, \dot{u}, \ldots, u^{(k)}, y, \dot{y}, \ldots, y^{(k)}\right) = 0, \quad i = 1, \ldots, p \qquad (5.18)$$

The conditions and the complete algorithm for rewriting the state-space model (3.19) into the form (5.18) can be found in [21]. The following simple example will illustrate the method without going too much into the details.

Example 5.2.2 (Nonlinear input–output model)
Rewriting a simple nonlinear state-space model into input–output form

Consider a simple nonlinear process model describing the operation of a continuous fermenter which has been introduced in Subsection 4.5.3:

$$\dot{x}_1 = \mu(x_2)x_1 - \frac{x_1}{V}u \qquad (5.19)$$

$$\dot{x}_2 = -\frac{\mu(x_2)x_1}{Y} + \frac{S_F - x_2}{V}u \qquad (5.20)$$

$$y = x_2 \qquad (5.21)$$

where $\mu(x_2) = \frac{x_2}{K+x_2}$ and Y, V and S_F are constants.

To express the direct relationship between u and y, x_1 and $\dot{x}_1$ should be eliminated from Equations (5.19)–(5.20). First, let us express x_1 from (5.20) using the output equation $y = x_2$:

$$x_1 = \frac{Y}{\mu(y)} \left(\frac{(S_F - y)u}{V} - \dot{y} \right) \tag{5.22}$$

Taking the time derivative of (5.22) gives

$$\dot{x}_1 = \frac{-Y}{y^2 V} \left(K S_F \dot{y} u - K V \dot{y}^2 + u y^2 \dot{y} - K S_F y \dot{u} \right.$$
$$\left. + K y^2 \dot{u} + K V y \ddot{y} - S_F y^2 \dot{u} + y^3 \dot{u} + V y^2 \ddot{y} \right) \tag{5.23}$$

Substituting (5.22) and (5.23) back into (5.19) gives the required (and quite complicated) time domain input–output relation of the model.

5.3 Realization Theory

The realization problem for a given input–output representation is to compute and characterize the properties of state-space representations that correspond to a specified linear or nonlinear system. In this sense, the realization problem is the reverse of the problem described in Subsection 5.1.3.

5.3.1 Realization of LTI Systems

Firstly, the realization problem of an SISO system from a given transfer function $H(s)$ is considered. The notations and the line followed in this subsection are based on [57]. The form of linear state equations we want to obtain is

$$\begin{aligned} \dot{x} &= Ax + Bu, \quad x(0) = 0 \\ y &= Cx \end{aligned} \tag{5.24}$$

where $x \in \mathbb{R}^n$, $A \in \mathbb{R}^{n \times n}$, $B \in \mathbb{R}^{n \times 1}$ and $C \in \mathbb{R}^{1 \times n}$. Of course, the model (5.24) must satisfy $H(s) = C(SI - A)^{-1}B$. The unknowns in the problem are the dimension of the state-space n and the matrices A, B and C. Of course, we are interested in finding a state-space model with minimal state-space dimension n. We know from Subsection 3.2.2 that state-space models are not unique, therefore we cannot expect a unique solution of the realization problem.

The term $(sI - A)^{-1}$ can be expanded into a negative power series in the following way:

$$(sI - A)^{-1} = Is^{-1} + As^{-2} + A^2 s^{-3} + \dots$$

therefore

$$C(sI - A)^{-1}B = CBs^{-1} + CABs^{-2} + CA^2Bs^{-3} + \dots$$

where the coefficients $h_i = CA^{i-1}B$, $i = 1, 2, \dots$ are the so-called *Markov parameters*. From this, it's clear that the realization problem for LTI systems is to find a state-space representation of the transfer functions of the form

$$H(s) = h_1 s^{-1} + h_2 s^{-2} + h_3 s^{-3} + \dots$$

To state the realization problem first in a more abstract way, let us define three operators.

Let $Z(s)$ be a negative power series of the form

$$Z(s) = k_1 s^{-1} + k_2 s^{-2} + k_3 s^{-3} + \dots \tag{5.25}$$

A *shift operator* $\mathbf{S}$ is defined as

$$\mathbf{S}Z(s) = k_2 s^{-1} + k_3 s^{-2} + k_4 s^{-3} + \dots \tag{5.26}$$

i.e. $\mathbf{S}$ shifts the coefficients of a power series one position to the left, and the left-most coefficient is dropped. The repeated (k-times) application of the shift operator is denoted by $\mathbf{S}^k$.

Using the shift operator and the transfer function $H(s)$, we can specify a linear space of negative power series over the real numbers as

$$U = \text{span}\{H(s), \mathbf{S}H(s), \mathbf{S}^2 H(s), \dots\} \tag{5.27}$$

It can be seen that the shift operator is an operator on U, $\mathbf{S}: U \mapsto U$.

We define the *initialization operator* $\mathbf{L}: \mathbb{R} \mapsto U$ as

$$\mathbf{L}r = H(s)r \tag{5.28}$$

The *evaluation operator* $\mathbf{E}: U \mapsto \mathbb{R}$ applied to a negative power series of the form (5.25) is defined as

$$\mathbf{E}Z(s) = k_1 \tag{5.29}$$

It is possible to show that the linear operators $\mathbf{S}$, $\mathbf{L}$ and $\mathbf{E}$ form a so-called *abstract realization* on the linear space U. This is often called the *shift realization* and it is denoted by $(\mathbf{S}, \mathbf{L}, \mathbf{E}, U)$.

The composition of $\mathbf{L}$, $\mathbf{S}$ and $\mathbf{E}$ gives the Markov parameters:

$$\mathbf{E}\mathbf{S}^0\mathbf{L} = \mathbf{E}H(s) = h_1$$
$$\mathbf{E}\mathbf{S}\mathbf{L} = \mathbf{E}\mathbf{S}H(s) = h_2$$
$$\mathbf{E}\mathbf{S}^2\mathbf{L} = \mathbf{E}\mathbf{S}^2 H(s) = h_3$$

$$\vdots \tag{5.30}$$

The following theorem provides a necessary and sufficient condition for the realizability of a linear time-invariant system in abstract terms.

Theorem 5.3.1 (Realization of SISO LTI systems). *A SISO LTI system described by the transfer function $H(s)$ is realizable in the form (5.24) if and only if the vector space U is finite dimensional. Furthermore, if the system is realizable, then (S,L,E,U) is a minimal linear realization.*

Algorithm for Obtaining a Minimal Realization. The method for obtaining a minimal realization of the form (5.24) is the following. Suppose that $H(s)$ is linearly realizable and thus U has finite dimension n. Then it can be shown that the abstract vectors

$$H(S), \mathbf{S}H(s), \ldots, \mathbf{S}^{n-1}H(s)$$

form a basis in U. Let us now identify these vectors with the standard basis vectors $e_i \in \mathbb{R}^n$, $i = 1, \ldots, n$

$$e_1 = H(s), \ e_2 = \mathbf{S}H(s), \ \ldots, \ e_n = \mathbf{S}^{n-1}H(s) \tag{5.31}$$

and write $\mathbf{S}^n H(s)$ as a linear combination of the basis vectors:

$$\mathbf{S}^n H(s) = \sum_{i=1}^{n} r_i \mathbf{S}^{i-1} H(s) \tag{5.32}$$

Then the $n \times n$ matrix A represents the shift operator as follows:

$$A = \begin{bmatrix} 0 & 0 & \ldots & 0 & r_1 \\ 1 & 0 & \ldots & 0 & r_2 \\ 0 & 1 & \ldots & 0 & r_3 \\ \vdots & \vdots & \vdots & \vdots & \vdots \\ 0 & 0 & \ldots & 1 & r_n \end{bmatrix} \tag{5.33}$$

Since the initialization operator $\mathbf{L}$ maps from a one-dimensional vector space to an n-dimensional space, it can be represented by an $n \times 1$ vector. From the definition of the operator it is clear that this vector is

$$B = e_1 = \begin{bmatrix} 1 \\ 0 \\ \vdots \\ 0 \end{bmatrix} \tag{5.34}$$

The evaluation operator $\mathbf{E}$ maps from an n-dimensional space to a one-dimensional space, therefore it can be represented by a $1 \times n$ dimensional vector. It can be seen from (5.30) that $\mathbf{E}e_i = h_i$ for $i = 1, \ldots, n-1$. Thus, $\mathbf{E}$ can be given by

$$C = \begin{bmatrix} h_1 & h_2 & \ldots & h_n \end{bmatrix} \tag{5.35}$$

In conclusion, we have generated a state-space representation in the form of Equation (5.24) with the matrices (A, B, C) above.

Realization using Hankel matrices. In Section 5.1.3 it was shown that the Markov parameters of an LTI system model are realization-independent. Another useful realization-independent structure composed of Markov parameters is the Hankel matrix of an LTI system.

Definition 5.3.1 (Hankel matrix, LTI case)
The Hankel matrix of an LTI system G is an infinite matrix of the form

$$H_G = \begin{bmatrix} h_1 & h_2 & h_3 & \cdots \\ h_2 & h_3 & h_4 & \cdots \\ h_3 & h_4 & h_5 & \cdots \\ \vdots & \vdots & \vdots & \end{bmatrix} \tag{5.36}$$

Note that the Hankel matrix depends only on (A, B, C) in the form of the coefficients of $h_i = CA^{i-1}B$.

Theorem 5.3.2 (Realization of SISO linear systems using Hankel matrices). *A linear system described by the transfer function $H(s)$ is realizable in the form (5.24) if and only if its Hankel matrix H_G is of finite rank. Furthermore, the rank of H_G is the dimension of the minimal linear realization of $H(s)$.*

Finally, we give another necessary and sufficient realizability condition that corresponds to MIMO linear systems and which is the easiest to check in practice. For this, consider the linear state-space model

$$\begin{aligned} \dot{x} &= Ax + Bu \\ y &= Cx, \quad x(0) = 0 \end{aligned} \tag{5.37}$$

where $x \in \mathbb{R}^n$, $u \in \mathbb{R}^r$, $y \in \mathbb{R}^p$ and the matrices A, B and C are of appropriate dimensions. The transfer function of (5.37) is the $p \times r$ matrix

$$H(s) = C(sI - A)^{-1}B$$

Theorem 5.3.3 (Realization of MIMO linear systems). *Consider a linear system described by a $p \times r$ transfer function matrix $H(s)$. Then the system is realizable by a finite dimensional linear state equation of the form (5.37) if and only if each element $H_{ij}(s)$ of $H(s)$ is a strictly proper rational function.*

5.3.2 Realization Theory for Nonlinear Systems

The task of finding a state-space realization of a nonlinear system from its input–output model is described by the following **problem statement**: given a nonlinear input–output description of a system, determine a state-space realization of the form (3.19):

$$\begin{aligned} \dot{x}(t) &= f(x(t)) + \sum_{i=1}^{m} g_i(x(t))u_i(t) \\ y(t) &= h(x(t)) \end{aligned}$$

Definition 5.3.2 (Realization of a formal power series)

The set $\{g_0, \ldots, g_m, h, x^0\}$ with $g_0 = f$ is called a realization of a formal power series c.

The Hankel matrix and Hankel Rank of Nonlinear Systems. The Hankel matrix of a nonlinear system plays a central role in realization theory. Its definition is a non-trivial extension of that of LTI systems as is described in abstract terms below.

Definition 5.3.3 (Hankel matrix, nonlinear case)

Given a formal power series c as an input–output realization of a nonlinear system. An infinite matrix H_c, in which each block of p-rows of index $(i_r, \ldots, i_0)$ on the column of index $(j_k, \ldots, j_0)$ is exactly the coefficient

$$c(i_r \ldots i_0 \, j_k \ldots j_0)$$

of c, is called the Hankel matrix of the series c.

Definition 5.3.4 (Hankel rank)

The rank of the matrix H_c above is called the Hankel rank of the system.

Note that the Hankel rank is not necessarily finite because the Hankel matrix is an infinite matrix.

Construction of the Hankel Matrix for Nonlinear Systems. Before approaching the construction of Hankel matrices from a Fliess's series expansion, some useful notations are introduced.

In order to enumerate the elements of the nonlinear Hankel matrix in a systematic way, we need to index its elements with the elements of I^* as follows:

1. An infinitely long column vector is indexed as:

index	element
(0)	a
(1)	b
(00)	c
(01)	d
(10)	e
(11)	f
...	...

   $a, b, c, d, e, f, \ldots \in \mathbb{R}$

2. An infinitely long block-column vector consisting of p-blocks, that is of column vectors with p elements, is also indexed with the elements of I^*.

index	element
(0)	a_1
	a_2
	$\ldots$
	a_p
(1)	b_1
	b_2
	$\ldots$
	b_p
$\ldots$	$\ldots$

$a, b \in \mathbb{R}^p$

Now we can construct the Hankel matrix of a nonlinear system represented by its Fliess's series expansion (see Equation (5.2.1)) by indexing

- its columns with the elements of I^*, and
- its row-blocks of length p with the elements of I^*.

Then the construction rule applies to the following items:
  row-block index: $(i_r \ldots i_0)$,
  column index: $(j_k \ldots j_0)$,
  matrix element : $c(i_r \ldots i_0 j_k \ldots j_0)$
where $c(i_k \ldots i_0) = L_{g_{i0}} \ldots L_{g_{ik}} h(x_0)$ and $c(\emptyset) = h(x_0)$ are the coefficients of the Fliess's series expansion.

Let us consider a special case, a nonlinear system with one input and p outputs. Then the first few columns of the Hankel matrix are as follows:

Columns $\emptyset$, (0), (1), (00)

	$\emptyset$	(0)	(1)	(00)
$\emptyset$	$h_1(x_0)$	$L_f h_1(x_0)$	$L_g h_1(x_0)$	$L_f^2 h_1(x_0)$
	$\vdots$	$\vdots$	$\vdots$	$\vdots$
	$h_p(x_0)$	$L_f h_p(x_0)$	$L_g h_p(x_0)$	$L_f^2 h_p(x_0)$
(0)	$L_f h_1(x_0)$	$L_f^2 h_1(x_0)$	$L_g L_f h_1(x_0)$	$L_f^3 h_1(x_0)$
	$\vdots$	$\vdots$	$\vdots$	$\vdots$
	$L_f h_p(x_0)$	$L_f^2 h_p(x_0)$	$L_g L_f h_p(x_0)$	$L_f^3 h_p(x_0)$
(1)	$L_g h_1(x_0)$	$L_f L_g h_1(x_0)$	$L_g^2 h_1(x_0)$	$L_f^2 L_g h_1(x_0)$
	$\vdots$	$\vdots$	$\vdots$	$\vdots$
	$L_g h_p(x_0)$	$L_f L_g h_p(x_0)$	$L_g^2 h_p(x_0)$	$L_f^2 L_g h_p(x_0)$
(00)	$L_f^2 h_1(x_0)$	$L_f^3 h_1(x_0)$	$L_g L_f^2 h_1(x_0)$	$L_f^4 h_1(x_0)$
	$\vdots$	$\vdots$	$\vdots$	$\vdots$
	$L_f^2 h_p(x_0)$	$L_f^3 h_p(x_0)$	$L_g L_f^2 h_p(x_0)$	$L_f^4 h_p(x_0)$
$\vdots$	$\vdots$	$\vdots$	$\vdots$	$\vdots$

Columns (01), (10), (11)

	(01)	(10)	(11)
$\emptyset$	$L_g L_f h_1(x_0)$	$L_f L_g h_1(x_0)$	$L_g^2 h_1(x_0)$
	$\vdots$	$\vdots$	$\vdots$
(0)	$L_g L_f h_p(x_0)$ $\cdots$	$L_f L_g h_p(x_0)$ $\cdots$	$L_g^2 h_p(x_0)$ $\cdots$
	$\vdots$	$\vdots$	$\vdots$
(1)	$\cdots$ $\cdots$	$\cdots$ $\cdots$	$\cdots$ $\cdots$
	$\vdots$	$\vdots$	$\vdots$
(00)	$\cdots$ $\cdots$	$\cdots$ $\cdots$	$\cdots$ $\cdots$
	$\vdots$	$\vdots$	$\vdots$
$\vdots$	$\cdots$ $\vdots$	$\cdots$ $\vdots$	$\cdots$ $\vdots$

5.3.3 Realization of Bilinear Systems

As we have already seen in Section 3.5.3, bilinear systems are special, easy-to-handle types of nonlinear input-affine systems. This is also true in realization theory: special strong results apply to bilinear systems which are the subject of this subsection.

The following example shows the structure of a Hankel matrix of bilinear systems.

Example 5.3.1 (Hankel matrix of a bilinear system)
Construction of the Hankel matrix for a simple SISO bilinear system

Consider first a simple special SISO bilinear state-space model in the form:

$$\dot{x} = Ax + Nxu, \quad y = h(x) = Cx$$

and let us introduce the following notation: $N_0 = A$, $N_1 = N$. Then the coefficients of the Fliess's series expansion as an input–output model of the system are in the following simple form:

$$c(\emptyset) = Cx_0$$
$$c(i_k \ldots i_0) = C N_{i_k} \ldots N_{i_0} x_0$$

Finally, the Hankel matrix of the system is as follows:

	$\emptyset$	(0)	(1)	$\cdots$
$\emptyset$	Cx_0	CAx_0	CNx_0	$\cdots$
(0)	CAx_0	CA^2x_0	$CANx_0$	$\cdots$
(1)	CNx_0	$CNAx_0$	CN^2x_0	$\cdots$
$\vdots$	$\vdots$	$\vdots$	$\vdots$	$\vdots$

The following theorem shows the importance of finite Hankel rank and bilinear system representations at the same time.

Theorem 5.3.4 (Existence of bilinear realization). *Let c be a formal power series in $m + 1$ non-commutative indeterminates and coefficients in $\mathbb{R}^p$. There exists a bilinear realization of c if and only if the Hankel rank of c is finite.*

5.4 Hankel Matrix of a 2-input–2-output Bilinear Heat Exchanger Cell model

As we have already seen in Section 4.4.4, the state-space model of a heat exchanger cell derived from first engineering principles is bilinear. We use this fact in this section to illustrate how to construct the Hankel matrix of a bilinear process system from its state-space representation.

Consider the dynamic model of a simple heat exchanger:

$$\frac{dT_{co}}{dt} = \frac{v_c}{V_c}(T_{ci} - T_{co}) + \frac{UA}{c_{pc}\rho_c V_c}(T_{ho} - T_{co}) \tag{5.38}$$

$$\frac{dT_{ho}}{dt} = \frac{v_h}{V_h}(T_{hi} - T_{ho}) + \frac{UA}{c_{ph}\rho_h V_h}(T_{co} - T_{ho}) \tag{5.39}$$

Let us introduce the following notations:

$$k_c = \frac{UA}{c_{pc}\rho_c V_c}, \quad k_h = \frac{UA}{c_{ph}\rho_h V_h} \tag{5.40}$$

$$x_1 = T_{co}, \; x_2 = T_{ho}, \; u_1 = v_c, \; u_2 = v_h \tag{5.41}$$

$$x = \begin{bmatrix} x_1 \\ x_2 \end{bmatrix}, \quad u = \begin{bmatrix} u_1 \\ u_2 \end{bmatrix} \tag{5.42}$$

With the notations above, the model can be written in the standard form

$$\dot{x} = Ax + Bu + \sum_{i=1}^{2} N_i x u_i \tag{5.43}$$

where

$$A = \begin{bmatrix} -k_c & k_c \\ k_h & -k_h \end{bmatrix} \tag{5.44}$$

$$B = \begin{bmatrix} \frac{T_{ci}}{V_c} & 0 \\ 0 & \frac{T_{hi}}{V_h} \end{bmatrix} \tag{5.45}$$

$$N_1 = \begin{bmatrix} \frac{-1}{V_c} & 0 \\ 0 & 0 \end{bmatrix} \tag{5.46}$$

$$N_2 = \begin{bmatrix} 0 & 0 \\ 0 & \frac{-1}{V_h} \end{bmatrix} \tag{5.47}$$

$$\tag{5.48}$$

Furthermore, let us define the output of the system as

$$y = \begin{bmatrix} y_1 \\ y_2 \end{bmatrix} = Cx = \begin{bmatrix} 1 & 0 \\ 0 & 1 \end{bmatrix} \begin{bmatrix} x_1 \\ x_2 \end{bmatrix} \tag{5.49}$$

Let us denote the column vectors of B by b_1 and b_2, and the row vectors of C by c_1 and c_2 respectively.

Based on the notation above, the scalar elements in the upper-left corner of the Hankel matrix of the heat exchanger with the defined inputs and outputs can be written as

	$\emptyset$	(0)	(1)	(2)
$\emptyset$	$c_1 x_0$	$c_1 A x_0$	$c_1(b_1 + N_1 x_0)$	$c_1(b_2 + N_2 x_0)$
	$c_2 x_0$	$c_2 A x_0$	$c_2(b_1 + N_1 x_0)$	$c_2(b_2 + N_2 x_0)$
(0)	$c_1 A x_0$	$c_1 A^2 x_0$	$c_1 A(b_1 + N_1 x_0)$	$c_1 A(b_2 + N_2 x_0)$
	$c_2 A x_0$	$c_2 A^2 x_0$	$c_2 A(b_1 + N_1 x_0)$	$c_2 A(b_2 + N_2 x_0)$
(1)	$c_1(b_1 + N_1 x_0)$	$c_1 N_1 A x_0$	$c_1 N_1(b_1 + N_1 x_0)$	$c_1 N_1(b_2 + N_2 x_0)$
	$c_2(b_1 + N_1 x_0)$	$c_2 N_1 A x_0$	$c_2 N_1(b_1 + N_1 x_0)$	$c_2 N_1(b_2 + N_2 x_0)$
(2)	$c_1(b_2 + N_2 x_0)$	$c_1 N_2 A x_0$	$c_1 N_2(b_1 + N_1 x_0)$	$c_1 N_2(b_2 + N_2 x_0)$
	$c_2(b_2 + N_2 x_0)$	$c_2 N_2 A x_0$	$c_2 N_2(b_1 + N_1 x_0)$	$c_2 N_2(b_2 + N_2 x_0)$

5.5 The Zero Dynamics

The zero dynamics is an important concept that plays a role exactly similar to the zeros of the transfer function in a linear system. The notion of zero dynamics was introduced by Byrnes and Isidori [17]. Its application to the solution of critical problems of asymptotic stabilization was described in [18].

The abstract definition of zero dynamics for input-affine nonlinear systems is as follows.

Definition 5.5.1 (Zero dynamics)
Consider the system (3.19) with constraints $y = 0$, that is

$$\begin{aligned} \dot{x} &= f(x) + \sum_{i=1}^{m} g_i(x) u_i \\ 0 &= h(x) \end{aligned} \tag{5.50}$$

The constrained system (5.50) is called the zero-output constrained dynamics, or briefly, the zero dynamics.

5.5.1 The Zero Dynamics of SISO Nonlinear Systems

Before we turn to analyzing the zero dynamics of SISO nonlinear systems, we introduce the notion of relative degree below.

Definition 5.5.2 (Relative degree)
The single-input–single-output nonlinear system,

$$\dot{x} = f(x) + g(x)u \tag{5.51}$$
$$y = h(x) \tag{5.52}$$

is said to have relative degree r at a point x^0 if

1. $L_g L_f^k h(x) = 0$ for all x in a neighborhood of x^0 and all $k < r - 1$.
2. $L_g L_f^{r-1} h(x^0) \neq 0$.

The last condition is called a non-triviality condition. Note that there may be points where a relative degree cannot be defined.

Note that the notion of relative degree will be extensively used later in Section 10.1 for designing nonlinear controllers using state feedback and input–output linearization.

After a suitable coordinates transformation (see, e.g. [37]) $z = \Phi(x)$ where $z_i = \phi_i(x) = L_f^{i-1} h(x)$ for $1 \leq i \leq r$ and $L_g \phi_i(x) = 0$ for $r + 1 \leq j \leq n$, the state-space model (3.19) with $m = 1$, $p = 1$ and relative degree r can be rewritten as

$$\dot{z}_1 = z_2$$
$$\dot{z}_2 = z_3$$
$$\cdots$$
$$\dot{z}_{r-1} = z_r$$
$$\dot{z}_r = b(\xi, \eta) + a(\xi, \eta)u$$
$$\dot{\eta} = q(\xi, \eta) \tag{5.53}$$

where $\xi = [z_1 \ \cdots \ z_r]^T$, $\eta = [z_{r+1} \ \cdots \ z_n]^T$, $a(\xi, \eta) = L_g L_f^{r-1} h(\Phi^{-1}(\xi, \eta))$ and $b(\xi, \eta) = L_f^r h(\Phi^{-1}(\xi, \eta))$.

Problem of Zeroing the Output. This problem is to find, if it exists, pairs consisting of an initial state x^* and input function u defined for all t in a neighborhood of $t = 0$, such that the corresponding output $y(t)$ of the system is identically zero for all t in a neighborhood of $t = 0$. For any fixed initial state x^*, the input function u can be determined as follows. Let us set the output to be identically zero, then the system's behavior is governed by the differential equation

$$\dot{\eta}(t) = q(0, \eta(t)) \tag{5.54}$$

The dynamics (5.54) describes the internal behavior of the system when the output is forced to be zero. The initial state of the system must be set to a value such that $\xi(0) = 0$, while $\eta(0) = \eta^0$ can be chosen arbitrarily. Furthermore, the input must be set as

$$u(t) = -\frac{b(0, \eta(t))}{a(0, \eta(t))} \tag{5.55}$$

where $\eta(t)$ denotes the solution of (5.54) with initial condition $\eta(0) = \eta^0$.

The investigation of the zero dynamics can be extremely useful when selecting the outputs to be controlled, since *the stabilization of an output with globally asymptotically stable zero dynamics implies the global asymptotic stability of the closed-loop system.*

Motivated by the notions of linear systems' theory, nonlinear systems with globally asymptotically stable zero dynamics are called *minimum-phase* systems.

5.5.2 Example: The Zero Dynamics of Continuous Fermentation Processes

The procedure of obtaining zero dynamics of a system will be illustrated in the example of a bio-reactor here. In order to analyze zero dynamics as is described in Section 5.5 before, we need to extend the original nonlinear two-dimensional state equation in Equations (4.78)–(4.79) with a nonlinear output equation

$$y = h(x) \tag{5.56}$$

where y is the output variable and h is a given nonlinear function. Then the zero dynamics of an input-affine nonlinear system containing two state variables can be analyzed using a suitable nonlinear coordinates transformation $z = \Phi(x)$:

$$\begin{bmatrix} z_1 \\ z_2 \end{bmatrix} = \begin{bmatrix} y \\ \lambda(x) \end{bmatrix} \tag{5.57}$$

where $\lambda(x)$ is a solution of the following partial differential equation (PDE):

$$L_g \lambda(x) = 0 \tag{5.58}$$

where $L_g \lambda(x) = \frac{\partial \lambda}{\partial x} g(x)$, i.e.

$$\frac{\partial \lambda}{\partial x_1} g_1 + \frac{\partial \lambda}{\partial x_2} g_2 = 0 \tag{5.59}$$

In the case of the simple fermenter model in Equations (4.78)–(4.79), we can analytically solve the above equation to obtain

$$\lambda(x) = F\left(\frac{V(-S_f + x_2 + S_0)}{x_1 + X_0}\right) \tag{5.60}$$

where F is an arbitrary continuously differentiable function. Then we can use the simplest possible coordinates transformation $z = \Phi(x)$ in the following form:

$$\begin{bmatrix} z_1 \\ z_2 \end{bmatrix} = \begin{bmatrix} y \\ \frac{V(-S_f + x_2 + S_0)}{x_1 + X_0} \end{bmatrix} \tag{5.61}$$

Selecting the Substrate Concentration as Output. If a linear function of the substrate concentration is chosen as output, *i.e.*

$$z_1 = y = k_s x_2 \tag{5.62}$$

where k_s is an arbitrary positive constant, then the inverse transformation $x = \Phi^{-1}(z)$ is given by

$$\begin{bmatrix} x_1 \\ x_2 \end{bmatrix} = \begin{bmatrix} \frac{-z_2 X_0 k_s - S_f V k_s + V z_1 + S_0 V k_s}{z_2 k_s} \\ \frac{z_1}{k_s} \end{bmatrix} \tag{5.63}$$

Thus the zero dynamics in the transformed coordinates can be computed as

$$\dot{z}_2 = \dot{\lambda} = L_f \lambda(x) = \frac{\partial \lambda}{\partial x} \dot{x} \tag{5.64}$$

which gives

$$\dot{z}_2 = L_f \lambda(x) + L_g \lambda(x) u = L_f \lambda(x) = L_f \lambda(\Phi^{-1}(z)) \tag{5.65}$$

since by construction $L_g \lambda(x) = 0$ (see Equation 5.59). The above equation is constrained by $y = k_s x_2 = z_1 = 0$. Then the *zero dynamics of the system is* given by the differential equation

$$\dot{z}_2 = L_f \lambda(\Phi^{-1}(0, z_2)) = -\frac{(z_2 Y + V) S_0 \mu_{max}}{Y(K_2 + S_0^2 + S_0 + K_1)} \tag{5.66}$$

which is *linear* and *globally stable*. The equilibrium state of the zero dynamics is at $z_2 = -\frac{V}{Y}$, which together with $z_1 = 0$ corresponds to the desired equilibrium state $x_1 = 0, x_2 = 0$ in the original coordinates. The above analysis shows that *if we manage to stabilize the substrate concentration either by a full state feedback or by an output feedback (partial state feedback) or even by a dynamic controller (which does not belong to the scope of this chapter) then the overall system will be stable.*

Selecting the Biomass Concentration as Output. The output in this case is a linear function of the biomass concentration:

$$z_1 = y = k_x x_1 \tag{5.67}$$

The zero dynamics of the system is given by:

$$\dot{z}_2 = -\frac{V\mu_{max}(z_2^2 Y X_0 + z_2(YVS_f + VX_0) + S_f V^2))}{(K_2 X_0^2 z_2^2 + z_2(2K_2 X_0 S_f V + VX_0) + V^2(K_2 S_f^2 + S_f + K_1))Y} \tag{5.68}$$

which describes a nonlinear dynamics and is only locally stable around the desired equilibrium state. The stability region can be determined by using the parameters of the system.

The right-hand side of (5.68) is visible in Figure 5.2 using the parameter values of Table 4.1. It can be seen that the zero dynamics has two equilibrium points: the equilibrium point corresponding to the optimal operating point ($z_2 = -V/Y = -8$) is stable and the other one is unstable.

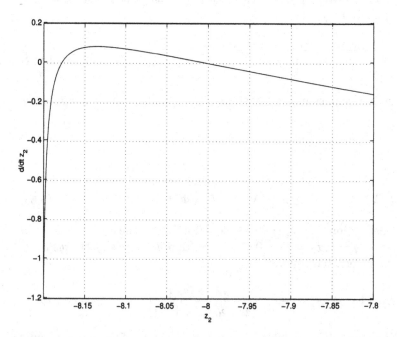

Figure 5.2. The zero dynamics in the transformed coordinates: continuous bioreactor, input: inlet feed flow rate, output: biomass concentration

The results of the analysis of the zero dynamics show that the best choice of output to be controlled is the substrate concentration and involving the

biomass concentration into the output generally brings singular points into the zero dynamics and makes the stability region narrower.

5.6 Further Reading

An excellent introduction to the theory of linear systems including realization theory can be found in [20]. The Hankel matrix approach to linear systems is emphasized in [56]. The realization problem of time-varying linear systems is examined in, *e.g.* [62].

In [25], the Fliess series expansions are discussed in detail. The basic reference for the Volterra series expansion of nonlinear systems is [45]. The existence and the relationship between the minimal realizations of nonlinear systems are investigated in [71]. The properties and input–output description of bilinear systems are studied in [16]. The theory of Hankel matrices for nonlinear systems is worked out in [24]. The problem of obtaining the time-domain input–output description of a nonlinear state- space model by eliminating the state variables is discussed in [79].

5.7 Summary

The input–output representation forms of lumped parameter systems in both the LTI and the nonlinear input-affine cases are described. The nonlinear input–output models are introduced as generalizations of the LTI forms in the form of formal power series and higher-order nonlinear differential equations.

The basic notions and main results of realization theory are also presented in both the LTI and the nonlinear input-affine cases., where one has to find a state-space model for a given input–output representation. Special emphasis has been put on certain system invariants: to Markov parameters and to Hankel matrices.

The notion of zero dynamics is also introduced in this chapter and it is illustrated in the example of a continuous fermentation process.

5.8 Questions and Application Exercises

Exercise 5.8.1. A continuous time LTI system is given with the following state-space representation:

$$\dot{x} = \begin{bmatrix} -1 & 2 \\ 1 & 0 \end{bmatrix} x + \begin{bmatrix} 1 \\ 0 \end{bmatrix} u$$

$$y = \begin{bmatrix} 1 & 2 \end{bmatrix} x$$

with the initial condition $x(0) = \begin{bmatrix} 0 \\ 0 \end{bmatrix}$

1. Give the transfer function of the system. Is the transfer function unique?
2. Is the state-space representation unique?
3. Compute the impulse-response function of the system.
4. Compute the response of the system (y) for the unit step input.

Exercise 5.8.2. A system is described by the following input–output model:

$$\dot{y}_2 = u_3$$
$$\dot{y}_3 + y_1 = u_2$$
$$\ddot{y}_3 + 2y_1 - y_3 = 8u_2 - 2u_1 + u_3 + 4u_4$$
$$2\dot{y}_1 + \dot{y}_3 + y_2 = u_4 - u_1$$

1. Give a possible state-space representation of the system.
2. Is the system linear and time-invariant? Why?
3. How many inputs, outputs, states has the system?
4. Is this state-space representation unique? Why?

Exercise 5.8.3. Show that the nonlinear state-space model

$$\dot{x}_1 = x_1^3 x_2 + u$$
$$\dot{x}_2 = -x_1 + \ln(x_2)$$
$$y = x_2, \quad x_2 > 0$$

is a realization of the input–output model

$$\dot{y} - y\ddot{y} = (\ln(y) - \dot{y})^3 y^2 + uy, \quad y > 0$$

Exercise 5.8.4. Examine the stability of the zero dynamics of the following system:

$$\dot{x}_1 = 2x_1 - x_2^2$$
$$\dot{x}_2 = -x_2^3 - x_1$$
$$y = x_1$$

Is any coordinates transformation needed for this?

Exercise 5.8.5. Calculate the terms in the Fliess series of the system in Exercise 5.8.4 for $x(0) = [0 \ 0]^T$ up to $k = 2$.

Exercise 5.8.6. Determine the upper 3×3 part of the Hankel matrix of the system in Exercise 5.8.4 for $x(0) = [0 \ 0]^T$.

6. Controllability and Observability

Controllability and observability play a fundamental role in designing controllers for both linear and nonlinear systems. This chapter is devoted to the basic as well as to the advanced material on controllability and observability presented in the following sections:

- *Controllability and observability of LTI systems*
  This section summarizes the basic notions of controllability and observability in the example of the most simple linear time-invariant system.
- *Local controllability and observability of nonlinear systems*
  The controllability and observability notions are extended here to the case of the input-affine nonlinear system.
- *Controllability and observability of nonlinear process systems*
  The local controllability and observability analysis is specialized in this section to the case of the nonlinear process system with no source.
- *Process system examples and case studies*
  Complete case studies illustrate the use of linear and nonlinear controllability and observability analysis methods on different heat exchanger models and on continuous and fed-batch fermenter models.

6.1 Controllability and Observability of LTI Systems

This section is devoted to the basic reference case, to the controllability and observability analysis of LTI systems, which will be extended to the nonlinear case later in this chapter.

Given an LTI system with finite dimensional representations in the form

$$\dot{x} = Ax + Bu$$
$$y = Cx$$
(6.1)

It is important to observe that the above general form of the state-space representation used for the controllability and observability analysis is a special case of the one in Chapter 3 in Equation (3.1) with $D = 0$. The reason for this simplification is that one always can transform the more general form in Equation (3.1) into Equation (6.1) by appropriate re-scaling (centering) of the input and output variables.

Thus a state-space representation will be characterized by the triplet (A, B, C) in this section. The following dynamical properties of state-space representations of LTI systems are described here:

1. Observability.
2. Controllability.

6.1.1 State Observability

In order to understand the problem statement of state observability, we need to recall from Chapter 2 that only the input and output variables of a system are directly observable (measurable) and not its state variables.

Definition 6.1.1 (State observability)
Given the inputs and the outputs of a system over a finite time interval. If it is possible to determine the value of the states based on these values and a state-space system model in such a way that we use functions of inputs and outputs and their derivatives, then the system is called (state) observable.

In the case of LTI systems, the above general definition specializes to the one below.

Definition 6.1.2 (LTI state observability)
Given a state-space model by its realization matrices (A, B, C) and the measured input and output signals

$$\{ u(t), \ y(t) \mid t_0 \ \leq t \leq t_F \}$$

The system is state observable if we can determine the state signal x at a given time t_0, i.e. $x(t_0)$.

The following theorem gives necessary and sufficient condition for a state-space representation of an LTI system to be state observable:

Theorem 6.1.1 (LTI state observability). *Given a state-space model of a LTI system by its realization matrices (A, B, C). This state-space model is state observable if and only if the observability matrix $\mathcal{O}_n$ is of full rank*

$$\mathcal{O}_n = \begin{bmatrix} C \\ CA \\ \cdot \\ \cdot \\ \cdot \\ CA^{n-1} \end{bmatrix}$$

where $\dim x = n$, *i.e.* $\operatorname{rank} \mathcal{O}_n = n$.

It is important to emphasize that *the observability matrix depends on the realization*, and it may change if one applies state transformation to a state-space representation (A, B, C). This holds because the observability matrix $\mathcal{O}_n$ depends only on the pair (A, C) in a way that is not invariant under state transformation. At the same time, the rank of the observability matrix $\mathcal{O}_n$ remains unchanged.

6.1.2 State Controllability

As its name suggests, state controllability is a necessary dynamic property of a system to be controllable in the entire state-space.

Definition 6.1.3 (State controllability)
A state-space model of a system is called state controllable if it is possible to drive any state $x(t_1)$ to any other state $x(t_2) \neq x(t_1)$ with an appropriate input in finite $t = t_2 - t_1$ time.

In the LTI case, the above definition specializes to the one below.

Definition 6.1.4 (LTI state controllability)
A state-space model of an LTI system given by its realization matrices (A, B, C) is called state controllable if it is possible to drive any state $x(t_1)$ to any other state $x(t_2) \neq x(t_1)$ with an appropriate input in finite $t = t_2 - t_1$ time.

The following theorem establishes necessary and sufficient condition of an LTI system to be state controllable in terms of its controllability matrix.

Theorem 6.1.2. *A state-space model of an LTI system with realization matrices (A, B, C) is state controllable if and only if the controllability matrix*

$$\mathcal{C}_n = \begin{bmatrix} B \ AB \ A^2B \ .. \ A^{n-1}B \end{bmatrix}$$

is of full rank, that is rank $\mathcal{C}_n = n$.

Again, it is important to emphasize that *the controllability matrix depends on the realization*, and it may change if one applies state transformation to a state-space representation (A, B, C). This holds because the controllability matrix $\mathcal{C}_n$ depends only on the pair (A, B) in a way that is not invariant under state transformation. However, the rank of the controllability matrix $\mathcal{C}_n$ remains unchanged.

The following simple example illustrates the notions and tools introduced above:

Example 6.1.1 (LTI controllability and observability)
Observability and controllability of a simple LTI state-space model

Investigate the controllability and observability of the following continuous time LTI MIMO system:

$$\dot{x} = \begin{bmatrix} -5 & -2 \\ 2 & 1 \end{bmatrix} x + \begin{bmatrix} 3 & 2 & -4 \\ 1 & 0 & 6 \end{bmatrix} u$$

$$y = \begin{bmatrix} 4 & 1 \\ -3 & 0 \end{bmatrix} x$$

The MIMO system has the following controllability matrix:

$$\mathcal{C} = \begin{bmatrix} B & AB \end{bmatrix} = \begin{bmatrix} 3 & 2 & -4 & -17 & -10 & 8 \\ 1 & 0 & 6 & 7 & 4 & -2 \end{bmatrix},$$

with rank $\mathcal{C} = 2 = n$, therefore the system is controllable. The observability matrix

$$\mathcal{O} = \begin{bmatrix} C \\ CA \end{bmatrix} = \begin{bmatrix} 4 & 1 \\ -3 & 0 \\ -18 & -7 \\ 15 & 6 \end{bmatrix}$$

has rank $\mathcal{O} = 2 = n$, so the system is observable.

6.1.3 Conditions for Joint Controllability and Observability

We recall that both the controllability and observability matrix changes if we apply state transformation to a given realization. This section will show that *joint* controllability and observability is a system property, that is, it is realization-independent.

Equivalent necessary and sufficient conditions will be given for an LTI system to be jointly controllable and observable. This will lead to the notion of *minimal realization* that will play an important role for nonlinear systems as well.

Joint Controllability and Observability is a System Property. The main result in this subsection uses the notion and properties of Hankel matrices introduced in Section 5.3.1.

The following lemma states that *joint* controllability and observability remains unchanged if a state transformation is applied, therefore it is a realization-independent property for LTI systems:

Lemma 6.1.1. *If we have an n-th order realization* (A, B, C) *of an LTI system with transfer function* $H(s) = \frac{b(s)}{a(s)}$ *which is controllable and observable, then all other n-th order realizations are controllable and observable.*

The proof uses the fact that one can construct a Hankel matrix from the observability and controllability matrices

$$\mathcal{O}(C,A) = \begin{bmatrix} C \\ CA \\ \cdot \\ \cdot \\ \cdot \\ CA^{n-1} \end{bmatrix}, \quad \mathcal{C}(A,B) = \begin{bmatrix} B\ AB\ A^2B\ \ldots\ A^{n-1}B \end{bmatrix}$$

where $\mathcal{O}(C,A)$ is a column block-matrix and $\mathcal{C}(A,B)$ is a row block-matrix.
    With the matrices above

$$H[1, n-1] = \mathcal{O}(C,A)\mathcal{C}(A,B) \tag{6.2}$$

The Irreducibility of the Transfer Function. An equivalent (that is necessary and sufficient) condition for an LTI system to be jointly controllable and observable is to have its transfer function to be irreducible.

Definition 6.1.5 (Irreducible transfer function)
The transfer function

$$H(s) = \frac{b(s)}{a(s)}$$

is called irreducible if the polynomials $a(s)$ and $b(s)$ are relative primes, that is, they have no common factors (no common roots).

Theorem 6.1.3. $H(s) = \frac{b(s)}{a(s)}$ *is irreducible if and only if all n-th order realizations are jointly controllable and observable.*

Minimal Realizations. The second equivalent condition to joint controllability and observability uses the notion of minimal realizations.

Definition 6.1.6 (Minimal realization)
A realization (A,B,C) of an LTI system of dimension $\dim x = n$ is minimal if one cannot find another realization of dimension less than n.

With this notion we can state an equivalence between minimality of a realization and irreducibility of the transfer function of the system.

Theorem 6.1.4. $H(s) = \frac{b(s)}{a(s)}$ *is irreducible if and only if any of its realizations (A,B,C) are minimal where*

$$H(s) = C(sI - A)^{-1}B$$

Using Theorem 6.1.3, we can now state the equivalence between minimality of a realization and joint controllability and observability of a system.

Theorem 6.1.5. *A realization (A,B,C) is minimal if and only if the system is jointly controllable and observable.*

It is important to note that minimal realizations are not unique, that is, we can find infinitely many minimal realizations which are related.

Lemma 6.1.2. *Any two minimal realizations can be connected by a unique similarity transformation (which is invertible).*

Let the two minimal realizations be (A_1, B_1, C_1) and (A_2, B_2, C_2). Minimal realizations are jointly controllable and observable, therefore the matrix

$$T = \mathcal{O}^{-1}(C_1, A_1)\mathcal{O}(C_2, A_2) = \mathcal{C}(A_1, B_1)\mathcal{C}^{-1}(A_2, B_2) \qquad (6.3)$$

exists and it is invertible.

Remember that any invertible similarity transformation T will produce another realization $(\overline{A}, \overline{B}, \overline{C})$ from the given realization (A, B, C) of the same system with the same transfer function, thus T serves as the transformation matrix between the two realizations.

The following simple example illustrates the investigation of joint controllability and observability of LTI systems:

Example 6.1.2 (LTI joint controllability and observability)
Joint controllability and observability of a simple LTI system

Let the simple LTI system be described by the following input–output model:

$$2y''' + 8y'' + 2y' - 12y = 4u + 2u'$$

Investigate joint controllability and observability of this system and give a state-space realization.

Applying Laplace transformation, we get

$$(2s^3 + 8s^2 + 2s - 12)Y(s) = (4 + 2s)U(s)$$

- From this, the transfer function can be computed:

$$H(s) = \frac{Y(s)}{U(s)} = \frac{2s + 4}{2s^3 + 8s^2 + 2s - 12} = \frac{s + 2}{s^3 + 4s^2 + s - 6}$$

- The root of the enumerator ($s_1 = -2$) is also a root of the denominator, so the transfer function is not irreducible. Therefore the system is not jointly controllable and observable.

- By using the controller form realization (see in Subsection 3.2.3), we can easily determine a state-space representation:

$$\dot{x} = \begin{bmatrix} -4 & -1 & 6 \\ 1 & 0 & 0 \\ 0 & 1 & 0 \end{bmatrix} x + \begin{bmatrix} 1 \\ 0 \\ 0 \end{bmatrix} u$$

$$y = \begin{bmatrix} 0 & 1 & 2 \end{bmatrix} x$$

6.1.4 General Decomposition Theorem

If we have a representation (A, B, C), then we can always find an invertible similarity transformation which moves the system to another realization $(\overline{A}, \overline{B}, \overline{C})$ with the partitioned state vector

$$\overline{x} = \begin{bmatrix} \overline{x}_{co} & \overline{x}_{c\overline{o}} & \overline{x}_{\overline{c}o} & \overline{x}_{\overline{c}\overline{o}} \end{bmatrix}^T$$

and partitioned matrices

$$\overline{A} = \begin{bmatrix} \overline{A}_{co} & 0 & \overline{A}_{13} & 0 \\ \overline{A}_{21} & \overline{A}_{c\overline{o}} & \overline{A}_{23} & \overline{A}_{24} \\ 0 & 0 & \overline{A}_{\overline{c}o} & 0 \\ 0 & 0 & \overline{A}_{43} & \overline{A}_{\overline{c}\overline{o}} \end{bmatrix} \qquad \overline{B} = \begin{bmatrix} \overline{B}_{co} \\ \overline{B}_{c\overline{o}} \\ 0 \\ 0 \end{bmatrix} \tag{6.4}$$

$$\overline{C} = \begin{bmatrix} \overline{C}_{co} & 0 & \overline{C}_{\overline{c}o} & 0 \end{bmatrix}$$

The partitioning above defines *subsystems* of the original system which are as follows:

Controllable and Observable Subsystem. The realization of the jointly controllable and observable subsystem is given by the matrices $\overline{A}_{co}, \overline{B}_{co}, \overline{C}_{co}$. The realization $(\overline{A}_{co}, \overline{B}_{co}, \overline{C}_{co})$ is minimal, *i.e.* $\overline{n} \leq n$ and

$$\overline{C}(s\overline{I} - \overline{A})^{-1}\overline{B} = C(sI - A)^{-1}B \tag{6.5}$$

Controllable Subsystem. Its realization is

$$\left(\begin{bmatrix} \overline{A}_{co} & 0 \\ \overline{A}_{21} & \overline{A}_{c\overline{o}} \end{bmatrix}, \begin{bmatrix} \overline{B}_{co} \\ \overline{B}_{c\overline{o}} \end{bmatrix}, \begin{bmatrix} \overline{C}_{co} & 0 \end{bmatrix} \right) \tag{6.6}$$

Observable Subsystem. Its realization is

$$\left(\begin{bmatrix} \overline{A}_{co} & \overline{A}_{13} \\ 0 & \overline{A}_{\overline{c}o} \end{bmatrix}, \begin{bmatrix} \overline{B}_{co} \\ 0 \end{bmatrix}, \begin{bmatrix} \overline{C}_{co} & \overline{C}_{\overline{c}o} \end{bmatrix} \right) \tag{6.7}$$

Uncontrollable and Unobservable Subsystem. Its realization is

$$([\overline{A_{\overline{co}}}], \quad [0], \quad [0]) \tag{6.8}$$

6.2 Local Controllability and Observability of Nonlinear Systems

This section is devoted to the extension of controllability and observability to the class of input-affine nonlinear systems given by a state-space representation in the form

$$\dot{x} = f(x) + g(x)u = f(x) + \sum_{i=1}^{m} g_i(x)u_i \tag{6.9}$$

$$y_j = h_j(x), \quad j = 1, \ldots, p \tag{6.10}$$

6.2.1 The Controllability Distribution, Controllable Nonlinear Systems

In the case of nonlinear systems, the set of states that are reachable from a given initial state are characterized using *distributions* (see Section A.4.1 in the Appendix).

The first results presented here are generalizations of the General Decomposition Theorem above. Their complete derivation can be found in [37].

Lemma 6.2.1. *Let Δ be a nonsingular involutive distribution of dimension d and suppose that Δ is invariant under the vector field f in Equation (6.9). Then at each point x^0 there exist a neighborhood U^0 of x^0 and a coordinates transformation $z = \Phi(x)$ defined on U^0, in which the vector field f is represented by a vector of the form*

$$\bar{f}(z) = \begin{bmatrix} \bar{f}_1(z_1, \ldots, z_d, z_{d+1}, \ldots, z_n) \\ \cdots \\ \bar{f}_d(z_1, \ldots, z_d, z_{d+1}, \ldots, z_n) \\ \bar{f}_{d+1}(z_{d+1}, \ldots, z_n) \\ \cdots \\ \bar{f}_n(z_{d+1}, \ldots, z_n) \end{bmatrix} \tag{6.11}$$

It is not difficult to prove that the last $n - d$ coordinate functions of the local coordinates transformation $z = \Phi(x)$ in the neighborhood of x^0 can be calculated from the following equality:

$$\text{span}\{d\Phi_{d+1}, \ldots, d\Phi_n\} = \Delta^{\perp} \tag{6.12}$$

while the first d coordinate functions should be chosen so that Φ is locally invertible around x^0 (*i.e.* the Jacobian of Φ evaluated at x^0 should be non-singular).

Now we can use the above Lemma to state the main result.

Proposition 6.2.1. *Let Δ be a nonsingular involutive distribution of dimension d and assume that Δ is invariant under the vector fields $f, g_1, \ldots, g_m$ in Equation (6.9). Moreover, suppose that the distribution $\text{span}\{g_1, \ldots, g_m\}$ is contained in Δ. Then, for each point x^0 it is possible to find a neighborhood U^0 of x^0 and a local coordinate transformation $z = \Phi(x)$ defined on U^0 such that, in the new coordinates, the system (6.9)–(6.10) is represented by equations of the form*

$$\dot{\zeta}_1 = f_1(\zeta_1, \zeta_2) + \sum_{i=1}^{m} g_{1i}(\zeta_1, \zeta_2) u_i \qquad (6.13)$$

$$\dot{\zeta}_2 = f_2(\zeta_2) \qquad (6.14)$$

$$y_i = h_i(\zeta_1, \zeta_2) \qquad (6.15)$$

where $\zeta_1 = (z_1, \ldots, z_d)$ and $\zeta_2 = (z_{d+1}, \ldots, z_n)$.

The above proposition (which presents a coordinate-dependent nonlinear analog of the controllability part of the general decomposition theorem for LTI systems) is very useful for understanding the input-state behavior of nonlinear systems.

Supposing that the assumptions of Proposition 6.2.1 are satisfied, choose a point x^0 and set $x(0) = x^0$. For small values of t, the state remains in U^0 and we can use Equations (6.13)–(6.15) to interpret the behavior of the system. From these, we can see that the ζ_2 coordinates of $x(t)$ are not affected by the input. If we denote by $x^0(T)$ the point of U^0 reached at time $t = T$ then it's clear that the set of points that can be reached at time T, starting from x^0, is a set of points whose ζ_2 coordinates are necessarily equal to the ζ_2 coordinates of $x^0(T)$. Roughly speaking, if we can find an appropriate Δ distribution and the local coordinates transformation $z = \Phi(x)$ then we can clearly identify the part of the system that behaves independently of the input in a neighborhood of x^0.

It is also important to note that if the dimension of Δ is equal to n then the dimension of the vector ζ_2 is 0, which means that the input affects all the state variables in a neighborhood of x^0 (the system is reachable in a neighborhood of x^0).

On the basis of Proposition 6.2.1 and the above explanation, the first step towards the analysis of local reachability of nonlinear systems is to find a distribution Δ_c that characterizes the controllability (reachability) of an input-affine nonlinear system.

Definition 6.2.1 (Controllability distribution)
A distribution Δ_c is called the controllability distribution of an input-affine nonlinear system if it possesses the following properties. It

1. is involutive, i.e.

$$\forall (\tau_1, \tau_2 \in \Delta_c) \quad \Longrightarrow \quad [\tau_1, \tau_2] \in \Delta_c$$

2. *is invariant under the vector fields* $(f = g_0, (g_i, \ i = 1, \ldots, m))$, *i.e.*

$$\forall(\tau \in \Delta_c) \quad => \quad [g_i, \tau] \in \Delta_c$$

3. *contains the distribution* $\mathrm{span}\{g_1, \ldots, g_m\}$

$$\Delta_0 = \mathrm{span}\{g_1, \ldots, g_m\} \subseteq \Delta_c$$

4. *is "minimal" (If D is a family of distributions on U, the smallest or minimal element is defined as the member of D (if it exists) which is contained in every other element of D) with the above properties.*

Lemma 6.2.2. *Let Δ be a given smooth distribution and $\tau_1, \ldots, \tau_q$ a given set of vector fields. The family of all distributions which are invariant under $\tau_1, \ldots, \tau_q$ and contain Δ has a minimal element, which is a smooth distribution.*

One can use the nonlinear analogue of the A-invariant subspace algorithm to construct the controllability distribution Δ_c.

Let us introduce the following *notation*. The smallest distribution that contains Δ and is invariant under the vector fields $g_0, \ldots, g_m$ will be denoted by $\langle g_0, \ldots, g_m | \Delta \rangle$.

Isidori [37] proposes an algorithm for constructing the controllability distribution as follows:

Algorithm for Constructing the Controllability Distribution.

1. *Starting point*

$$\Delta_0 = \mathrm{span}\{g_1, \ldots, g_m\} \tag{6.16}$$

2. *Development of the controllability distribution*

$$\Delta_k = \Delta_{k-1} + \sum_{i=0}^{m} [g_i, \Delta_{k-1}] \tag{6.17}$$

Note that one term in the last sum $[g_i, \Delta_{k-1}]$ is computed by using the functions $(\phi_1, \ldots, \phi_\ell)$ spanning the distribution Δ_{k-1}:

$$[g_i, \Delta_{k-1}] = \mathrm{span}\{[g_i, \phi_1], \ldots, [g_i, \phi_\ell]\}$$

It is proved that Δ_k has the property

$$\Delta_k \subset \langle g_0, \ldots, g_m | \Delta_0 \rangle$$

3. *Stopping condition*
   If $\exists k^*$ such that $\Delta_{k^*} = \Delta_{k^*+1}$, then

$$\Delta_c = \Delta_{k^*} = \langle g_0, \ldots, g_m | \Delta_0 \rangle$$

Properties of the Algorithm. The algorithm above exhibits some interesting properties. It starts with the distribution spanned by the input functions $g_i(x)$ of the original state equation. Thereafter it is necessary to compute the Lie-brackets (*i.e.* $[f(x), g_i(x)]$) of the functions $f(x)$ and $g_i(x)$ respectively. Then we expand the distribution obtained in the previous step by the distribution spanned by the Lie-brackets, *i.e.* $([f(x), g_i(x)], \ i = 1, \ldots, m_k)$.

Simple Examples. We start with the example of computing the controllability distribution of LTI systems, which highlights the connection between the A-invariant subspace algorithm and the algorithm for constructing the nonlinear controllability distribution.

Example 6.2.1 (LTI controllability distribution)
Controllability distribution of LTI systems

The above concepts of constructing the controllability distribution are complete analogues of the linear system concepts (infimal A-invariant subspaces). To see this, consider the linear system:

$$\dot{x} = Ax + Bu, \qquad u \in \mathbb{R}^m, \qquad x \in \mathbb{R}^n \tag{6.18}$$

$$y = Cx \tag{6.19}$$

Let us construct the smallest A-invariant subspace over Im B. Let

$$\Delta_0 = \text{Im } B = \text{span}\{b_1, \ldots, b_m\}$$

where b_i is the i-th column of matrix B and

$$f(x) = g_0(x) = Ax$$

at each $x \in \mathbb{R}^n$. Any vector field of Δ_0 can be expressed as

$$\theta(x) = \sum_{i=1}^{m} c_i(x) b_i$$

Then the above algorithm for constructing the controllability distribution works as follows:

$$\Delta_0 = \text{span}\{b_1, \ldots, b_m\}$$

Using the algorithm for Δ_k:

$$\Delta_1 = \Delta_0 + \sum_{i=0}^{m} [g_i, \Delta_0] = \text{span}\{b_1, \ldots, b_m, [f, b_1], \ldots, [f, b_m]\}$$

---

since

$$[g_i, g_j] = [b_i, b_j] = L_{b_i} b_j - L_{b_j} b_i = 0$$

Therefore with

$$[f, b_i](x) = [Ax, b_i] = \frac{\partial b_i}{\partial x^T} Ax - \frac{\partial (Ax)}{\partial x^T} b_i = -Ab_i, \quad i = 1, \ldots, m$$

we obtain that

$$\Delta_1 = \text{Im} \left[B | AB \right]$$

Similarly:

$$\Delta_k = \text{Im} \left[B | AB | \ldots | A^k B \right]$$

The following simple example illustrates the computation of the controllability distribution in a two-dimensional case.

Example 6.2.2 (Controllability distribution)
Controllability distribution of a simple two-dimensional system

Given a simple input-affine nonlinear system model in the form:

$$\dot{x}_1 = -x_1 e^{-\frac{1}{x_1}} + 3e^{x_2} - x_1 u$$
$$\dot{x}_2 = 5x_1 e^{-\frac{1}{x_1}} - x_2 u \qquad\qquad (6.20)$$
$$y = -x_2$$

First we extract the functions f and g from the system model above:

$$f(x_1, x_2) = \begin{pmatrix} -x_1 e^{-\frac{1}{x_1}} + 3e^{x_2} \\ 5x_1 e^{-\frac{1}{x_1}} \end{pmatrix}, \quad g(x_1, x_2) = \begin{pmatrix} -x_1 \\ -x_2 \end{pmatrix}$$

There are only two steps needed for computing the controllability distributions as follows:

$$\Delta_0 = \text{span}\{ g \}, \quad \Delta_1 = \text{span}\{ g, [f, g] \}$$

where $[f, g]$ is the Lie-product of f and g:

$$[f, g](x) = \frac{\partial g}{\partial x} f(x) - \frac{\partial f}{\partial x} g(x)$$

$$= \begin{bmatrix} -1 & 0 \\ 0 & -1 \end{bmatrix} \begin{pmatrix} -x_1 e^{-\frac{1}{x_1}} + 3e^{x_2} \\ 5x_1 e^{-\frac{1}{x_1}} \end{pmatrix} - \begin{bmatrix} -(\frac{x_1+1}{x_1})e^{-\frac{1}{x_1}} & 3e^{x_2} \\ 5(\frac{x_1+1}{x_1})e^{-\frac{1}{x_1}} & 0 \end{bmatrix} \begin{pmatrix} -x_1 \\ -x_2 \end{pmatrix}$$

$$= \begin{pmatrix} -e^{-\frac{1}{x_1}} + 3(x_2 - 1)e^{x_2} \\ 5x_1 e^{-\frac{1}{x_1}} \end{pmatrix}$$

Thus the controllability distribution is

$$\Delta(x) = \Delta_1(x) = \text{span} \left\{ \begin{pmatrix} -x_1 \\ -x_2 \end{pmatrix}, \begin{pmatrix} -e^{-\frac{1}{x_1}} + 3(x_2 - 1)e^{x_2} \\ 5x_1 e^{-\frac{1}{x_1}} \end{pmatrix} \right\}$$

Obtaining the Transformed System. The locally transformed (decomposed) system (6.13)–(6.15) can be calculated by performing the so-called *total integration* of the controllability distribution. The total integration basically means the solution of the set of quasi-linear partial differential equations (see (6.12)).

$$\frac{d\lambda_j}{dx} \left(f_1(x) \ldots f_d(x) \right) = 0 \tag{6.21}$$

for obtaining the functions λ_j, $j = 1, \ldots, n - d$, where the distribution to be integrated is spanned by the vector fields $f_1, \ldots, f_d$ and $f_i : \mathbb{R}^n \mapsto \mathbb{R}^n$ for $i = 1, \ldots, n$, $d < n$ and the $n - d$ λ functions are linearly independent.

According to the famous Frobenius theorem (see e.g. [37], Chapter 1), this problem is solvable if and only if the nonsingular distribution to be integrated is involutive. Note that the controllability distribution is always involutive by construction.

After solving the $n - d$ partial differential equations we can define the local coordinates transformation Φ by using the solution λ_j, $j = 1, \ldots, n - d$ as follows. Set the last $n - d$ coordinate functions of Φ as

$$\phi_{d+1}(x) = \lambda_1(x), \ldots, \phi_n(x) = \lambda_{n-d}(x) \tag{6.22}$$

Then, choose the first d coordinate functions from the coordinate functions of the identical mapping

$$x_1(x) = x_1, \ x_2(x) = x_2, \ldots, x_n(x) = x_n \tag{6.23}$$

such that the Jacobian of Φ is nonsingular (*i.e.* it is at least locally invertible) in the region of the state-space which is of interest.

An example of calculating the coordinate transformation Φ and totally integrating the controllability distribution of a nonlinear process system can be found later in Section 6.6.

6.2.2 The Observability Co-distribution, Observable Nonlinear Systems

Roughly speaking, the problem statement of observability in the nonlinear case is the following. Under what conditions can we distinguish the initial states of an input-affine nonlinear system described by Equations (6.9)–(6.10) by observing its outputs? We will examine this property locally, similarly to local controllability. For this, we need the definition of indistinguishable states and observability. The notations and results presented in this section are based on [21] and [37].

Let us denote the output of the system model (6.9)–(6.10) for input u and initial state $x(0) = x_0$ by $y(t, 0, x_0, u)$.

Definition 6.2.2 (Indistinguishable states, observable system)
Two states $x_1, x_2 \in X$ are called indistinguishable (denoted by $x_1 I x_2$) for (6.9)–(6.10) if for every admissible input function u the output functions $t \mapsto y(t, 0, x_1, u)$ and $t \mapsto y(t, 0, x_2, u)$, $t \geq 0$ are identical.
The system is called observable if $x_1 I x_2$ implies $x_1 = x_2$.

The local versions of the above properties are the following:

Definition 6.2.3 (V-indistinguishable states, local observability)
Let $V \subset X$ be an open set and $x_1, x_2 \in V$. The states x_1 and x_2 are said to be V-indistinguishable (denoted by $x_1 I^V x_2$), if for every admissible constant control u with the solutions $x(t, 0, x_1, u)$ and $x(t, 0, x_2, u)$ remaining in V for $t \leq T$, the output functions $y(t, 0, x_1, u)$ and $y(t, 0, x_2, u)$ are the same for $t \leq T$.

The system (3.19) is called locally observable at x_0 if there exists a neighborhood W of x_0 such that for every neighborhood $V \in W$ of x_0 the relation $x_0 I^V x_1$ implies $x_1 = x_0$. If the system is locally observable at each x_0 then it is called locally observable.

Definition 6.2.4 (Observation space)
The observation space $\mathcal{O}$ of the system (3.19) is the linear space of functions on X containing $h_1, \ldots, h_p$ and all repeated Lie-derivatives

$$L_{\tau_1} L_{\tau_2} \ldots L_{\tau_k} h_j, \quad j = 1, \ldots, p, \quad k = 1, 2, \ldots \tag{6.24}$$

where $\tau_i \in \{g_0, g_1, \ldots, g_m\}$, $i = 1, \ldots, k$.

We remark that the observation space has the interpretation that it contains the output functions and all of their derivatives along the system trajectories.

The following theorem gives a sufficient condition for local observability.

Theorem 6.2.1. *Consider the system (3.19) with $\dim X = n$ and assume that $\dim d\mathcal{O}(x_0) = n$ where*

$$d\mathcal{O}(x) = \text{span}\{dH(x) \mid H \in \mathcal{O}\}, \quad x \in X$$

Then the system is locally observable at x_0.

Based on this, it's useful to define the so-called observability co-distribution.

Definition 6.2.5 (Observability co-distribution)
The observability co-distribution $d\mathcal{O}$ of an input-affine nonlinear system with observation space $\mathcal{O}$ is defined as follows:

$$d\mathcal{O}(x) = \text{span}\{dH(x) \mid H \in \mathcal{O}\}, \quad x \in \mathcal{X} \tag{6.25}$$

The rank of the observability co-distribution can be determined using the dual version of the algorithm that was used for generating the controllability distribution.

Let us denote the smallest co-distribution which contains

$$\Omega = \text{span}\{dh_1, \ldots, dh_p\}$$

and is invariant under $g_0, \ldots, g_m$ by $\langle g_0, \ldots, g_m | \Omega \rangle$.

Algorithm for Constructing the Observability Co-distribution.

1. *Starting point*
$$\Omega_0 = \text{span}\{dh_1, \ldots, dh_p\}$$

2. *Developing the observability co-distribution*
$$\Omega_k = \Omega_{k-1} + \sum_{i=0}^{m} L_{g_i} \Omega_{k-1}$$

3. *Stopping criterion*
   If there exists an integer k^* such that $\Omega_{k^*} = \Omega_{k^*+1}$, then
$$\Omega_o = \Omega_{k^*} = \langle g_0, \ldots, g_m | \Omega_0 \rangle$$

The dimension of the nonsingular co-distribution Ω_{k^*} at x_0 is equal to the dimension of the observability co-distribution at x_0.

If the dimension of the observability co-distribution is strictly less than n, then we can find a local coordinates transformation which shows the unobservable nonlinear combinations of the state variables, as is stated by the following proposition:

Proposition 6.2.2. *Let Δ be a nonsingular involutive distribution of dimension d and assume that Δ is invariant under the vector fields $f, g_1, \ldots, g_m$. Moreover, suppose that the co-distribution $\text{span}\{dh_1, \ldots, dh_p\}$ is contained in the co-distribution $\Delta^\perp$. Then, for each point x_0 it is possible to find a neighborhood U^0 of x^0 and a local coordinates transformation $z = \Phi(x)$ defined on U^0 such that the system (6.9)–(6.10) is represented as*

$$\dot{\zeta}_1 = f_1(\zeta_1, \zeta_2) + \sum_{i=1}^{m} g_{1i}(\zeta_1, \zeta_2) u_i \tag{6.26}$$

$$\dot{\zeta}_2 = f_2(\zeta_2) + \sum_{i=1}^{m} g_{2i}(\zeta_2) u_i \tag{6.27}$$

$$y_i = h_i(\zeta_2) \tag{6.28}$$

where $\zeta_1 = (z_1, \ldots, z_d)$ and $\zeta_2 = (z_{d+1}, \ldots, z_n)$

It is evident from Equations (6.26)–(6.28) that the output depends only on ζ_2, and ζ_2 is independent of ζ_1. Therefore, starting from a fixed initial value of ζ_2 and from arbitrary initial values of ζ_1 and for arbitrary input u, the system produces exactly the same output and therefore it cannot be locally observable (see Definition 6.2.3).

It is important to note the duality and similarity between the algorithms generating a controllability distribution and the observability co-distribution.

Simple Examples. The following example shows the use of the algorithm of generating the observability co-distribution.

Example 6.2.3 (Observability co-distribution)

Consider again the system in Example 6.2.2 and let us calculate its observability co-distribution. The starting point of the algorithm is

$$\Omega_0(x) = \text{span}\{dh(x)\} = \text{span}\{[0 \; -1]\} \tag{6.29}$$

The Lie-product of $\omega = dh$ along f according to the definition is

$$L_f \omega(x) = f^T(x) \left(\frac{\partial \omega^T}{\omega x} \right)^T + \omega(x) \frac{\partial f}{\partial x} = \left[-5 \left(\frac{x_1 + 1}{x_1} \right) e^{\frac{1}{x_1}} \; 0 \right] \tag{6.30}$$

and

$$L_g \omega(x) = [0 \; 1] \tag{6.31}$$

Therefore the observability co-distribution after one step is given by

$$\Omega_1(x) = \text{span}\left\{ [0 - 1], \left[-5\left(\frac{x_1 + 1}{x_1} \right) e^{\frac{1}{x_1}} \quad 0 \right] \right\}, \tag{6.32}$$

from which we can see that the system satisfies local observability conditions at almost all points of the state-space.

In order to highlight the connection between linear and nonlinear observability, we construct the observability co-distribution of LTI systems in the next example.

Example 6.2.4 (LTI observability co-distribution)
Observability co-distribution of LTI systems

Consider the linear system

$$\dot{x} = Ax$$
$$y = Cx$$

Then

$$\Omega_0(x) = \text{span}\{c_1, \ldots, c_p\} \tag{6.33}$$
$$\tau(x) = Ax \tag{6.34}$$

where $c_1, \ldots, c_p$ denote the rows of C. The first step of the algorithm is

$$\Omega_1 = \Omega_0 + L_\tau \Omega_0 = \text{span}\{c_1, \ldots, c_p, L_\tau c_1, \ldots, L_\tau c_p\} \tag{6.35}$$

Since

$$L_\tau c_i(x) = L_{Ax} c_i = c_i \frac{\partial(Ax)}{\partial x} = c_i A \tag{6.36}$$

we have

$$\Omega_1(x) = \text{span}\{c_1, \ldots, c_p, c_1 A, \ldots, c_p A\} \tag{6.37}$$

Continuing in the same way, we have, for any $k \geq 1$,

$$\Omega_k(x) = \text{span}\{c_1, \ldots, c_p, c_1 A, \ldots, c_p A, \ldots, c_1 A^k, \ldots, c_p A^k\} \tag{6.38}$$

By duality, $\Omega_{n-1}^{\perp}$ is the largest distribution invariant under the vector field Ax and contained in the distribution $\Omega_0^{\perp}$. Note that at each $x \in \mathbb{R}^n$,

$$\Omega_0^{\perp}(x) = \ker(C) \tag{6.39}$$

$$\Omega_{n-1}^{\perp}(x) = \ker \begin{bmatrix} C \\ CA \\ \ldots \\ CA^{n-1} \end{bmatrix} \tag{6.40}$$

We are interested in the dimension of $\Omega_{n-1}^{\perp}(x)$ (which is independent of x). If the observability matrix is of full rank (n), then the dimension (d) of $\Omega_{n-1}^{\perp}(x)$ is 0, which means that the system is state observable.

6.2.3 The Minimal Realization of Nonlinear Systems

The solution to the problem of identifying the minimal realizations of a nonlinear system is similar to the linear case.

Theorem 6.2.2 (The minimal realization of nonlinear systems). *A realization $\{g_0, g_1, \ldots, g_m, h, x^0\}$ of a formal power series c is minimal if and only if the realization satisfies the controllability rank condition and the observability rank condition at x^0.*

6.3 Controllability and Observability of Nonlinear Process Systems

If one performs controllability analysis of process systems, the specialities of process system models should be taken into account. For this purpose the decomposed form of the general state equation (4.15) in Section 4.2.3 serves as a starting point. The nonlinear functions on the right-hand side of the input-affine state equations take the following special form:

$$f(x) = A_{transfer}x + Q_{\phi}^{(j)}(x) \tag{6.41}$$

$$g_i(x) = N_i x + B_{conv}^{(i)} \tag{6.42}$$

where $B_{conv}^{(i)}$ is the i-th column of the matrix B_{conv}.

We recall that the Lie-bracket is a multi-linear function of both of its arguments, therefore any of the components of the algorithm for constructing the controllability distribution has the following decomposition:

$$\Delta_0 = \text{span}\{N_1 x + B^{(1)}_{conv}, \ldots, N_m x + B^{(m)}_{conv}\} \tag{6.43}$$
$$\Delta_1 = \Delta_0 + \sum_{i=1}^{m} \text{span}\{(N_i A_{transfer} - A_{transfer} N_i)x$$
$$- A_{transfer} B^{(i)}_{conv} + [Q^{(j)}_\phi(x), N_i x] + \left[Q^{(j)}_\phi(x), B^{(i)}_{conv}\right]\}$$
$$+ \sum_{i=1}^{m} \sum_{k=1}^{m} \text{span}\{(N_k N_i - N_i N_k)x - N_i B^{(k)}_{conv} + N_k B^{(i)}_{conv}\}$$

The above decomposition shows that the controllability property is super-additive with respect to the mechanisms present in the terms of the conservation balance such as input and output convection, transfer and sources (including chemical reactions). This means that *two mechanisms both lacking controllability may interact to produce that property in the nonlinear case* [82].

6.3.1 Process Systems with no Source: the Linear-bilinear Time-invariant Case

In this special but still interesting case, there are no sources present in the system. Therefore the nonlinear functions of the general state equation in Equation (4.15) take the following decomposed special form:

$$f_L(x) = A_{transfer} x \tag{6.44}$$
$$g_i(x) = N_i x + B^{(i)}_{conv} \tag{6.45}$$

Observe that the state function is now linear and time-invariant and the input function remains bilinear-linear and time-invariant.

In this case the algorithm described above in Equation (6.43) for constructing the controllability distribution Δ_c is more simple because of the lack of the general nonlinear term and goes as follows:

$$\Delta_0 = \text{span}\{N_1 x + B^{(1)}_{conv}, \ldots, N_m x + B^{(m)}_{conv}\} \tag{6.46}$$
$$\Delta_1 = \Delta_0$$
$$+ \sum_{i=1}^{m} \text{span}\{(N_i A_{transfer} - A_{transfer} N_i)x - A_{transfer} B^{(i)}_{conv}\}$$
$$+ \sum_{i=1}^{m} \sum_{k=1}^{m} \text{span}\{(N_k N_i - N_i N_k)x - N_i B^{(k)}_{conv} + N_k B^{(i)}_{conv}\} \tag{6.47}$$

One can easily recognize the presence of the Kalman controllability matrix as an additive constant term in the vectors

$$A^k_{transfer} B^{(i)}, \quad k = 1, \ldots, n, \quad i = 1, \ldots, m$$

6.4 Heat Exchanger Examples

As we have seen earlier in Section 4.4, heat exchangers are one of the most simple yet important operating units in process systems. Depending on the modeling assumptions, heat exchangers may have LTI, LTV, LPV or input-affine bilinear state-space models, and thus they are ideal simple examples to carry out and compare linear and nonlinear controllability and observability analysis.

6.4.1 Local Controllability and Observability of an LTI Heat Exchanger Cell Model

As the most simple case, we start with the LTI state-space model of the heat exchanger cell developed in Section 4.4.2. It is relatively easy to construct the controllability distribution and the observability co-distribution of the heat exchanger cell in this case as follows:

- *Controllability distribution*
  The initial distribution

$$\Delta_0 = \text{span}\{g_1, g_2\} = \text{span}\{\begin{bmatrix} \frac{v_c}{V_c} \\ 0 \end{bmatrix}, \begin{bmatrix} 0 \\ \frac{v_h}{V_h} \end{bmatrix}\} \tag{6.48}$$

spans the whole $\mathbb{R}^2$, since its elements are constant non-zero functions that are linearly independent in any point. This means that the system is controllable by inlet temperatures at any point of the state-space.
- *Observability co-distribution*
  The initial co-distribution is given by the row vectors of the matrix C.

$$\Omega_0 = \text{span}\{[1 \ 0], [0 \ 1]\} \tag{6.49}$$

Its rank is always 2 so the system is observable at any point.

6.4.2 Local Controllability and Observability of a Nonlinear Heat Exchanger Cell

Next we consider the bilinear (nonlinear) input-affine state-space model of the heat exchanger cell developed in Section 4.4.4.

Now we need to use the algorithms above in Sections 6.2.1 and 6.2.2 for constructing the controllability distribution and the observability co-distribution of the nonlinear heat exchanger cell.

- *Controllability distribution*
  According to Equation (6.16) we can simply write the initial distribution Δ_0 as

$$\Delta_0(x) = \text{span}\{g_1(x), g_2(x)\} = \text{span}\{\begin{bmatrix} \frac{T_{ci}}{V_c} - \frac{1}{V_c}x_1 \\ 0 \end{bmatrix}, \begin{bmatrix} 0 \\ \frac{T_{hi}}{V_h} - \frac{1}{V_h}x_2 \end{bmatrix}\} \tag{6.50}$$

We can see from (6.50) that the subspace spanned by the functions g_1 and g_2 is two-dimensional for almost any x_1, x_2. The dimension only decreases to zero when $T_{ci} = x_1$ and $T_{hi} = x_2$ (*i.e.* the corresponding input and output temperatures are equal). This can happen if the cold and hot inlet temperatures are equal for a long time and thus there is no heat exchange in the operating unit, which is rarely the case in practice.
In the next step we compute the distribution Δ_1.

$$\Delta_1 = \Delta_0 + \text{span}\{[f, g_1],\ [f, g_2],\ [g_1, g_2]\} \tag{6.51}$$

First, we have to determine $[f, g_1]$.

$$[f(x), g_1(x)] = \frac{\partial g_1}{\partial x} f(x) - \frac{\partial f}{\partial x} g_1(x) \tag{6.52}$$

$$\frac{\partial g_1}{\partial x} f(x) = \begin{bmatrix} \frac{1}{V_c} & 0 \\ 0 & 0 \end{bmatrix} \begin{bmatrix} -k_1 x_1 + k_1 x_2 \\ k_2 x_1 - k_2 x_2 \end{bmatrix} = \begin{bmatrix} -\frac{1}{V_c}(-k_1 x_1 + k_1 x_2) \\ 0 \end{bmatrix} \tag{6.53}$$

$$\frac{\partial f}{\partial x} g_1(x) = \begin{bmatrix} k_1 & k_1 \\ k_2 & -k_2 \end{bmatrix} \begin{bmatrix} \frac{T_{ci}}{V_c} - \frac{1}{V_c} x_1 \\ 0 \end{bmatrix} = \begin{bmatrix} -k_1(\frac{T_{ci}}{V_c} - \frac{1}{V_c} x_1) \\ k_2(\frac{T_{ci}}{V_c} - \frac{1}{V_c} x_2) \end{bmatrix} \tag{6.54}$$

Thus

$$[f(x), g_1(x)] = \begin{bmatrix} -\frac{1}{V_c}(-k_1 x_1 + k_1 x_2) + k_1(\frac{T_{ci}}{V_c} - \frac{1}{V_c} x_1) \\ -k_2(\frac{T_{ci}}{V_c} - \frac{1}{V_c} x_2) \end{bmatrix} \tag{6.55}$$

After performing the same calculations for f and g_2 we get

$$[f(x), g_2(x)] = \begin{bmatrix} -k_1(\frac{T_{hi}}{V_h} - \frac{1}{V_h} x_2) \\ -\frac{1}{V_h}(-k_2 x_1 + k_2 x_2) + k_2(\frac{T_{hi}}{V_h} - \frac{1}{V_h} x_2) \end{bmatrix} \tag{6.56}$$

It's easy to compute that

$$[g_1(x), g_2(x)] = \begin{bmatrix} 0 \\ 0 \end{bmatrix} \tag{6.57}$$

Note that the above equation shows that the two input variables act on disjoint sets of state variables.
With the above equations, we have for Δ_1

$$\Delta_1 = \text{span}\{g_1, g_2, [f, g_1], [f, g_2], [g_1, g_2]\} \tag{6.58}$$

We can see from Equations (6.55) and (6.56) that the dimension of Δ_1 is still zero if the corresponding inlet and outlet temperatures and the hot and cold side outlet temperatures are equal respectively. That is, we could not improve Δ_0 and the algorithm stops here.
In conclusion, we can say that the nonlinear heat exchanger cell model is controllable with the exception of the singular point ($x_1 = T_{ci}$, $x_2 = T_{hi}$).

- *Observability co-distribution*

  Consider the nonlinear heat exchanger cell model (4.62)–(4.64) together with the output equation $y = x_1$. The observability co-distribution after the first step of the algorithm is

$$\text{span}\left\{\begin{bmatrix} 1 & 0 \end{bmatrix}, \begin{bmatrix} -k_1 & k_1 \end{bmatrix}\right\}$$

which is of rank 2 at any point of the state-space, therefore the nonlinear heat exchanger cell with the above output selection is observable in the nonlinear sense.

If we consider the *engineering meaning of the non-controllability condition* above, that is the corresponding inlet and outlet temperatures and the hot and cold side outlet temperatures are equal respectively, we can see that in this case there is no heat transfer between the hot and cold sides because they are at their equilibrium conditions. This means that the outlet temperatures of the heat exchanger will not change whatever happens to the flow rates, that is, the heat exchanger can't be controlled via changing the flow rates, which is in perfect agreement with the computations above.

6.4.3 Local Controllability of a Nonlinear 2-cell Heat Exchanger

Consider the bilinear (nonlinear) state-space model of the heat exchanger cell developed in Subsection 4.4.4. We can construct a simple model of a countercurrent heat exchanger by connecting two such cells. The nonlinear functions f and g on the right-hand side of the input-affine state-space model are now as follows:

$$f(x) = g_0(x) = \begin{bmatrix} -k_{1c}x_1 + k_{1c}x_2 \\ k_{1h}x_1 - k_{1h}x_2 \\ -k_{2c}x_3 + k_{2c}x_4 \\ k_{2h}x_3 - k_{2h}x_4 \end{bmatrix} \tag{6.59}$$

$$g_1(x) = \begin{bmatrix} V_{1c}^{-1}(x_3 - x_1) \\ 0 \\ V_{2c}^{-1}(T_{ci} - x_3) \\ 0 \end{bmatrix}, \quad g_2(x) = \begin{bmatrix} 0 \\ V_{1h}^{-1}(T_{hi} - x_2) \\ 0 \\ V_{2h}^{-1}(x_2 - x_4) \end{bmatrix} \tag{6.60}$$

Now we are in a position to compute the *controllability distribution* of the 2-cell heat exchanger system as follows:

$$\Delta_0 = \text{span}\{g_1, g_2\} \tag{6.61}$$

$$\Delta_1 = \text{span}\{g_1, g_2, [f, g_1], [f, g_2]\} \tag{6.62}$$

Similarly to the bilinear heat exchanger cell case, the necessary Lie-brackets are computed as follows:

$$\frac{\partial f(x)}{\partial x} = \begin{bmatrix} -k_{1c} & k_{1c} & 0 & 0 \\ k_{1h} & -k_{1h} & 0 & 0 \\ 0 & 0 & -k_{2c} & k_{2c} \\ 0 & 0 & k_{2h} & -k_{2h} \end{bmatrix} \tag{6.63}$$

$$\frac{\partial g_1(x)}{\partial x} = \begin{bmatrix} -V_{1c}^{-1} & 0 & V_{1c}^{-1} & 0 \\ 0 & 0 & 0 & 0 \\ 0 & 0 & -V_{2c}^{-1} & 0 \\ 0 & 0 & 0 & 0 \end{bmatrix}, \quad \frac{\partial g_2(x)}{\partial x} = \begin{bmatrix} 0 & 0 & 0 & 0 \\ 0 & -V_{1h}^{-1} & 0 & 0 \\ 0 & 0 & 0 & 0 \\ 0 & V_{2h}^{-1} & 0 & -V_{2h}^{-1} \end{bmatrix} \tag{6.64}$$

$$[g_1, g_2](x) = -[g_2, g_1](x) = \begin{bmatrix} 0 \\ 0 \\ 0 \\ 0 \end{bmatrix} \tag{6.65}$$

$$[f, g_1](x) = \frac{\partial g_1(x)}{\partial x} f(x) - \frac{\partial f(x)}{\partial x} g_1(x)$$

$$= \begin{bmatrix} -V_{1c}^{-1}(-k_{1c}x_1 + k_{1c}x_2) + V_{1c}^{-1}(-k_{2c}x_3 + k_{2c}x_4) \\ 0 \\ -V_{2c}^{-1}(-k_{2c}x_3 + k_{2c}x_4) \\ 0 \end{bmatrix}$$

$$- \begin{bmatrix} -k_{1c}V_{1c}^{-1}(x_3 - x_1) \\ k_{1h}V_{1c}^{-1}(x_3 - x_1) \\ -k_{2c}V_{2c}^{-1}(T_{ci} - x_3) \\ k_{2h}V_{2c}^{-1}(T_{ci} - x_3) \end{bmatrix}$$

$$= \begin{bmatrix} k_{1c}V_{1c}^{-1}(x_3 - x_2) + k_{2c}V_{1c}^{-1}(x_4 - x_3) \\ -k_{1h}V_{1c}^{-1}(x_3 - x_1) \\ k_{2c}V_{2c}^{-1}(T_{ci} - x_4) \\ -k_{2h}V_{2c}^{-1}(T_{ci} - x_3) \end{bmatrix} \tag{6.66}$$

$$[f, g_2](x) = \frac{\partial g_2(x)}{\partial x} f(x) - \frac{\partial f(x)}{\partial x} g_2(x)$$

$$= \begin{bmatrix} 0 \\ -V_{1h}^{-1}(k_{1h}x_1 - k_{1h}x_2) & 0 \\ V_{2h}^{-1}(k_{1h}x_1 - k_{1h}x_2) - V_{2h}^{-1}(k_{2h}x_3 - k_{2h}x_4) \end{bmatrix}$$

$$- \begin{bmatrix} k_{1c}V_{1h}^{-1}(T_{hi} - x_2) \\ -k_{1h}V_{1h}^{-1}(T_{hi} - x_2) \\ k_{2c}V_{2h}^{-1}(x_2 - x_4) \\ -k_{2h}V_{2h}^{-1}(x_2 - x_4) \end{bmatrix}$$

$$= \begin{bmatrix} -k_{1c}V_{1h}^{-1}(T_{hi} - x_2) \\ k_{1h}V_{1h}^{-1}(T_{hi} - x_1) \\ -k_{2c}V_{2h}^{-1}(x_2 - x_4) \\ k_{1h}V_{2h}^{-1}(x_1 - x_2) + k_{2h}V_{2h}^{-1}(x_2 - x_3) \end{bmatrix} \tag{6.67}$$

Controllability Analysis. We can see from Equations (6.61) and (6.60) that the subspace spanned by the functions g_1 and g_2 is four-dimensional for almost any x_1, ..., x_4. The computation of Δ_1 shows that the dimension of the distribution has not increased (it cannot in fact) therefore the bilinear 2-cell heat exchanger model is controllable everywhere with the exception of the singular points.

Singular Points. The dimension only decreases when

$$T_{ci} = x_1 \ , \ x_3 = x_1 \ , \ T_{hi} = x_2 \text{ or } x_2 = x_4$$

i.e. the corresponding input and output temperatures are equal in any of the cells on the hot or cold side. Similarly to the single heat exchanger cell case, this can only happen if the cold and hot inlet temperatures are equal for a long time and thus there is no heat exchange in the operating unit, which is rarely the case in practice.

6.5 Controllability of a Simple Continuous Fermentation Process

In this section it is shown in the example of a simple continuous fermenter described in Subsection 4.5.3 that nonlinear controllability analysis based on the generation of *controllability distributions* is extremely helpful in identifying the singular points of the state-space around which control of the system is problematic or even impossible [76].

The nonlinear state-space model of the fermenter is given in Equations (4.78)–(4.79). The variables and parameters of the fermentation process model are collected in Table 4.1.

6.5.1 Local Controllability Analysis Using the Linearized Model

The linearized state-space model of the continuous fermenter is also developed in Subsection 4.5.3 and it is given in Equations (4.84)–(4.87). After calculating the Kalman controllability matrix of the linearized model

$$\mathcal{O} = [B \ AB] = \begin{bmatrix} -1.2194 & 0 \\ 2.4388 & -0.0031 \end{bmatrix} \tag{6.68}$$

we find that the system is controllable (in the linear sense) in the neighborhood of the required operating point.

6.5.2 Nonlinear Controllability Analysis Using Controllability Distributions

For this, we need to identify the vector fields f and g in Equations (4.79) in Subsection 4.5.3. If we write them in the following general form

$$\dot{x} = f(x) + g(x)u \tag{6.69}$$

then it's clear that

$$x = \begin{bmatrix} x_1 \\ x_2 \end{bmatrix} = \begin{bmatrix} \bar{X} \\ \bar{S} \end{bmatrix}, \quad u = \bar{F} \tag{6.70}$$

and

$$f(x) = f(\bar{X}, \bar{S}) = \begin{bmatrix} \mu(\bar{S} + S_0)(\bar{X} + X_0) - \frac{(\bar{X}+X_0)F_0}{V} \\ -\frac{\mu(\bar{S}+S_0)(\bar{X}+X_0)}{Y} + \frac{(S_F-(\bar{S}+S_0))F_0}{V} \end{bmatrix} \tag{6.71}$$

$$g(x) = g(\bar{X}, \bar{S}) = \begin{bmatrix} -\frac{(\bar{X}+X_0)}{V} \\ \frac{(S_F-(\bar{S}+S_0))}{V} \end{bmatrix} \tag{6.72}$$

We follow the algorithm described in Subsection 6.2.1 for calculating the local controllability distribution of the model. The initial distribution is

$$\Delta_0 = \text{span}\{g\} \tag{6.73}$$

In the following step we add the Lie-bracket of f and g to the initial distribution

$$\Delta_1 = \Delta_0 + \text{span}\{[f,g]\} = \text{span}\{g, [f,g]\} \tag{6.74}$$

where

$$[f,g](x) = \frac{\partial g}{\partial x} f(x) - \frac{\partial f}{\partial x} g(x) \tag{6.75}$$

The second step then gives the following:

$$\Delta_2 = \Delta_1 + [f, \Delta_1] + [g, \Delta_1] = \text{span}\{g, [f,g], [f,[f,g]], [g,[f,g]]\} \tag{6.76}$$

Singular Points. At the point

$$[\bar{X} \ \bar{S}]^T = [-X_0 \ S_F - S_0]^T \quad \text{with} \quad X = 0 \begin{bmatrix} g \\ l \end{bmatrix}, S = S_F$$

all the elements of Δ_2 (and, of course, all the elements of Δ_0 and Δ_1) are equally zero. It means that the controllability distribution has rank 0 at this point. Moreover, this singular point is a steady-state point in the state-space. From this it follows that if the system reaches this (undesired) point, it's impossible to drive the process out of it by manipulating the input feed flow rate.

Δ_2 has rank 1 if

$$\bar{X} = -X_0 \ (X = 0 \begin{bmatrix} g \\ l \end{bmatrix}) \quad \text{and} \quad \bar{S} \neq S_F - S_0 \ (i.e. \ S \neq S_F)$$

From a practical point of view it means that if the biomass concentration decreases to $0 \begin{bmatrix} g \\ l \end{bmatrix}$ then it can't be increased by changing the input flow rate.

Stability analysis of these singular points show that both of them are stable.

Non-singular Points. At any other point in the state-space including the desired operating point $\begin{bmatrix} \bar{X} & \bar{S} \end{bmatrix}^T = \begin{bmatrix} 0 & 0 \end{bmatrix}^T$, the controllability distribution has rank 2, which means that the system is controllable in a neighborhood of these points and we can apply state feedback controllers to stabilize the process.

6.6 Controllability (Reachability) of Fed-batch Fermentation Processes

The controllability (reachability) of a simple nonlinear fed-batch fermentation process model is investigated in this section. It is shown that the known difficulties of controlling such processes are primarily caused by the fact that the rank of the controllability distribution is always less than the number of state variables.

Furthermore, a coordinates transformation is calculated analytically that shows the nonlinear combination of the state variables which is independent of the input. The results of the reachability analysis and that of the coordinates transformation are independent of the source function in the system model.

The results are extended to the four state variable non-isotherm case, and to nonlinear fed-batch chemical reactors with general reaction kinetics.

6.6.1 Problem Statement

Bio-processes in general and fermentation processes in particular are difficult to model and to control even in the simplest cases. Various difficulties are reported in the literature, which include instability and controllability problems for both continuous and fed-batch fermenters ([44], [43]).

The dynamic state-space model of a fermenter is derived from first engineering principles which fixes certain structural elements in the model. As we have already seen in Subsection 4.5.2, the state equations are derived from dynamic conservation balances of the overall mass, component masses and energy if applicable. The speciality of a fermentation model appears in the so-called source function of these balances, which is highly nonlinear and non-monotonous in nature.

The aim of this section is to use rigorous nonlinear analysis of a simple fed-batch fermenter model for analyzing its reachability (controllability) properties and to relate them to the physico-chemical phenomena taking place in the reactor [75].

6.6.2 Nonlinear State-space Model

The simplest dynamic model of a fed-batch fermenter has been developed in Subsection 4.5.2, which can be written in the following input-affine form [44]:

$$\dot{x} = f(x) + g(x)u \tag{6.77}$$

where

$$x = \begin{bmatrix} x_1 \\ x_2 \\ x_3 \end{bmatrix} = \begin{bmatrix} X \\ S \\ V \end{bmatrix}, \quad u = F \tag{6.78}$$

$$f(x) = \begin{bmatrix} \mu(x_2)x_1 \\ -\frac{1}{Y}\mu(x_2)x_1 \\ 0 \end{bmatrix} = \begin{bmatrix} \frac{\mu_{max}x_2 x_1}{K_1+x_2+K_2 x_2^2} \\ -\frac{\mu_{max}x_2 x_1}{(K_1+x_2+K_2 x_2^2)Y} \\ 0 \end{bmatrix}, \quad g(x) = \begin{bmatrix} -\frac{x_1}{x_3} \\ \frac{S_f-x_2}{x_3} \\ 1 \end{bmatrix} \tag{6.79}$$

and

$$\mu(x_2) = \frac{\mu_{max}x_2}{K_1 + x_2 + K_2 x_2^2} \tag{6.80}$$

Note that the above model is exactly the same as Equations (4.71)–(4.74) and they are repeated here for convenience.

The variables of the model with their units and the constant parameters with their typical values are also listed in Subsection 4.5.2.

6.6.3 Reachability Analysis

We construct the reachability distribution according to the algorithm described in Subsection 6.2.1 as follows:

$$\Delta_0 = \text{span}\{g\}$$

$$\Delta_1 = \Delta_0 + [f, \Delta_0] = \text{span}\{g, [f, g]\}$$

$$\Delta_2 = \Delta_1 + [f, \Delta_1] + [g, \Delta_1] = \text{span}\{g, [f, g], [f, [f, g]], [g, [f, g]]\}$$

The calculation of the Lie-products in Δ_1 and Δ_2 is as follows:

$$[f, g](x) = \frac{\partial g}{\partial x}f(x) - \frac{\partial f}{\partial x}g(x) \tag{6.81}$$

Since

$$\frac{\partial g}{\partial x}f(x) = \begin{bmatrix} -\frac{1}{x_3} & 0 & \frac{x_1}{x_3^2} \\ 0 & -\frac{1}{x_3} & \frac{x_2-S_f}{x_3^2} \\ 0 & 0 & 0 \end{bmatrix} \begin{bmatrix} f_1(x) \\ -\frac{1}{Y}f_1(x) \\ 0 \end{bmatrix} = \begin{bmatrix} -\frac{1}{x_3}f_1(x) \\ \frac{1}{Y}\frac{1}{x_3}f_1(x) \\ 0 \end{bmatrix} \tag{6.82}$$

and

$$\frac{\partial f}{\partial x}g(x) = \begin{bmatrix} \mu(x_2) & \frac{\partial \mu}{\partial x_2} & 0 \\ -\frac{1}{Y}\mu(x_2) & -\frac{1}{Y}\frac{\partial \mu}{\partial x_2} & 0 \\ 0 & 0 & 0 \end{bmatrix} \begin{bmatrix} g_1(x) \\ g_2(x) \\ 0 \end{bmatrix}$$

$$\begin{bmatrix} \mu(x_2)g_1(x) + \frac{\partial \mu}{\partial x_2}g_2(x) \\ -\frac{1}{Y}(\mu(x_2)g_1(x) + \frac{\partial \mu}{\partial x_2}g_2(x)) \\ 0 \end{bmatrix} \tag{6.83}$$

the Lie-product $[f, g]$ has the form

$$\begin{bmatrix} [f,g]_1 \\ [f,g]_2 \\ [f,g]_3 \end{bmatrix} = \begin{bmatrix} [f,g]_1 \\ -\frac{1}{Y}[f,g]_1 \\ 0 \end{bmatrix} \tag{6.84}$$

where $[f, g]_i$ denotes the i-th coordinate function of the vector field $[f, g]$.

It follows from Equations (6.81)–(6.84) that the distributions $[f, [f, g]]$ and $[g, [f, g]]$ will also have the same form as (6.84), i.e.

$$[f, [f, g]] = \begin{bmatrix} [f, [f, g]]_1 \\ -\frac{1}{Y}[f, [f, g]]_1 \\ 0 \end{bmatrix} \tag{6.85}$$

and

$$[g, [f, g]] = \begin{bmatrix} [g, [f, g]]_1 \\ -\frac{1}{Y}[g, [f, g]]_1 \\ 0 \end{bmatrix} \tag{6.86}$$

On the basis of the above we can denote the coordinate functions of the vector fields spanning Δ_2 at a given point x of the state-space as follows:

$$\Delta_2(x) = \text{span}\left\{ \begin{bmatrix} \delta_{11}(x) & \delta_{12}(x) & \delta_{13}(x) & \delta_{14}(x) \\ \delta_{21}(x) & \delta_{22}(x) & \delta_{23}(x) & \delta_{24}(x) \\ \delta_{31}(x) & \delta_{32}(x) & \delta_{33}(x) & \delta_{34}(x) \end{bmatrix} \right\} \tag{6.87}$$

where

$$\delta_{31} = 1, \quad \delta_{32}(x) = \delta_{33}(x) = \delta_{34}(x) = 0 \tag{6.88}$$

and

$$\delta_{22}(x) = -\frac{1}{Y}\delta_{12}(x) \tag{6.89}$$

$$\delta_{23}(x) = -\frac{1}{Y}\delta_{13}(x) \tag{6.90}$$

$$\delta_{24}(x) = -\frac{1}{Y}\delta_{14}(x) \tag{6.91}$$

i.e.

$$\Delta_2(x) = \text{span}\left\{ \begin{bmatrix} \delta_{11}(x) & \delta_{12}(x) & \delta_{13}(x) & \delta_{14}(x) \\ \delta_{21}(x) & -\frac{1}{Y}\delta_{12}(x) & -\frac{1}{Y}\delta_{13}(x) & -\frac{1}{Y}\delta_{14}(x) \\ 1 & 0 & 0 & 0 \end{bmatrix} \right\} \tag{6.92}$$

which means that we couldn't increase the dimension of the reachability distribution in the second step and the rank of Δ_2 is at most 2 at any point in the state-space.

Singular Points. There are, however, points in the state-space where the rank of the reachability distribution Δ_2 is less than 2. Those points are characterized by

$$x_1 = 0$$

In this case, Δ_2 is of dimension 1. This case means that there is no biomass in the system and since the inlet flow contains only substrate, the biomass concentration cannot be influenced by manipulating the input.

During the following analysis we will consider the open region of the state-space where Δ_1 is nonsingular and the value of state vector has real physical meaning (the concentrations and the liquid volume are positive), $i.e.$

$$U = \{x_1, x_2, x_3 | x_1 > 0, x_2 > 0, x_3 > 0\} \tag{6.93}$$

6.6.4 Calculation of the Coordinate Transformation

Since the generation of the reachability distribution stopped in the second step with

$$\Delta_1 = \mathrm{span}\{g, [f, g]\}$$

Δ_1 is the smallest distribution invariant under f, g and containing the vector field g. This distribution is denoted by $\langle f, g | \mathrm{span}\{g\}\rangle$. Since $\langle f, g | \mathrm{span}\{g\}\rangle$ is nonsingular on U and involutive we may use it to find a coordinates transformation $z = \Psi(x)$. The system in the new coordinates will be represented by equations of the following form (see Theorem 6.2.1):

$$\dot{\zeta}_1 = \bar{f}_1(\zeta_1, \zeta_2) + \bar{g}(\zeta_1, \zeta_2)u \tag{6.94}$$

$$\dot{\zeta}_2 = \bar{f}_2(\zeta_2) \tag{6.95}$$

where $\zeta_1 = [z_1 \; z_2]^T$ and $\zeta_2 = z_3$ in our case.

To calculate Φ, we have to integrate the distribution Δ_1 first, that is, to find a single $(\dim(x) - \dim(\Delta_1) = 3 - 2 = 1)$ real-valued function λ such that $\mathrm{span}\{d\lambda\} = [\langle f, g | \mathrm{span}\{g\}\rangle]^{\perp}$, where the sign $\perp$ denotes the annihilator of a distribution. Since

$$[f, g](x) = \begin{bmatrix} \dfrac{\left(\dfrac{\mu_{max}x_1}{K_1 + x_2 + K_2 x_2^2} - \dfrac{\mu_{max}x_2 x_1(1 + 2K_2 x_2)}{(K_1 + x_2 + K_2 x_2^2)^2}\right)(S_f - x_2)}{x_3} \\[3ex] \dfrac{\left(-\dfrac{\mu_{max}x_1}{(K_1 + x_2 + K_2 x_2^2)Y} + \dfrac{\mu_{max}x_2 x_1(1 + 2K_2 x_2)}{(K_1 + x_2 + K_2 x_2^2)^2 Y}\right)(S_f - x_2)}{x_3} \\[3ex] 0 \end{bmatrix} \tag{6.96}$$

this amounts to solve the partial differential equations (PDEs)

$$\begin{bmatrix} \dfrac{\partial\lambda}{\partial x_1} & \dfrac{\partial\lambda}{\partial x_2} & \dfrac{\partial\lambda}{\partial x_3} \end{bmatrix} \begin{bmatrix} -\dfrac{x_1}{x_3} & \dfrac{\left(\dfrac{\mu_{max}x_1}{K_1 + x_2 + K_2 x_2^2} - \dfrac{\mu_{max}x_2 x_1(1 + 2K_2 x_2)}{(K_1 + x_2 + K_2 x_2^2)^2}\right)(S_f - x_2)}{x_3} \\[3ex] \dfrac{S_f - x_2}{x_3} & \dfrac{\left(-\dfrac{\mu_{max}x_1}{(K_1 + x_2 + K_2 x_2^2)Y} + \dfrac{\mu_{max}x_2 x_1(1 + 2K_2 x_2)}{(K_1 + x_2 + K_2 x_2^2)^2 Y}\right)(S_f - x_2)}{x_3} \\[3ex] 1 & 0 \end{bmatrix}$$

$$= \begin{bmatrix} 0 & 0 \end{bmatrix} \tag{6.97}$$

Solution by the Method of Characteristics. The method of characteristics (see, *e.g.* [12], [8] or [53]) is used for solving the above resultant first-order linear homogeneous partial differential equation in the following general form:

$$\sum_{i=1}^{n} \phi_i(x)\partial_i\lambda(x) = 0, \quad \partial_i\lambda = \frac{\partial\lambda}{\partial x_i} \tag{6.98}$$

or briefly

$$\phi(x)\lambda'(x) = 0, \tag{6.99}$$

where $T \subset \mathbb{R}^n$ is a domain, $x \in T$, ϕ_i, $i = 1\ldots n$ are known functions and λ is the unknown. The characteristic equation system of (6.99) is the following set of ordinary differential equations:

$$\dot{\xi} = \phi(\xi) \tag{6.100}$$

We call the $\xi : \mathbb{R} \to \mathbb{R}^n$ solutions of (6.100) characteristic curves. A $\lambda \in C^1(T)$ function is called the first integral of (6.100) if $t \to \lambda(\xi(t))$ is constant along any characteristic curve. In order to solve (6.99) we have to find $(n-1)$ linearly independent solutions $(\lambda_1, \lambda_2, \ldots, \lambda_{n-1})$ of it. Then the general solution of (6.99) will be in the form $\lambda = \varPhi(\lambda_1, \lambda_2, \ldots, \lambda_{n-1})$, where $\varPhi \in C^1(\mathbb{R}^{n-1})$ is an arbitrary function. We know that a first integral of (6.100) satisfies (6.99), therefore we have to find $(n - 1)$ linearly independent first integrals to obtain the general solution. This can be done without solving (6.100), as is illustrated below in our case.

To solve the first PDE, namely

$$\frac{\partial\lambda}{\partial x_1}\left(-\frac{x_1}{x_3}\right) + \frac{\partial\lambda}{\partial x_2}\left(\frac{S_f - x_2}{x_3}\right) + \frac{\partial\lambda}{\partial x_2} = 0 \tag{6.101}$$

we start from the following set of ordinary differential equations:

$$\dot{x}_1 = -\frac{x_1}{x_3}$$

$$\dot{x}_2 = \frac{S_f - x_2}{x_3}$$

$$\dot{x}_3 = 1$$

It's easy to observe that

$$\dot{x}_1 x_3 = -x_1$$

and

$$\dot{x}_1 x_3 + \dot{x}_3 x_1 = (x_1 x_3)' = 0$$

since $\dot{x}_3 = 1$. Therefore $x_1 x_3 = \text{const.}$ Moreover,

$$\dot{x}_2 x_3 = S_f - x_2$$

and

$$\dot{x}_2 x_3 + x_2 \dot{x}_3 - S_f \dot{x}_3 = (x_2 x_3)' - S_f \dot{x}_3 = 0$$

from which it follows that

$$x_2 x_3 - S_f x_3 = \text{const.}$$

We can see from the above that the solution of (6.101) will be in the form

$$\lambda(x_1, x_2, x_3) = \Phi(x_1 x_3, x_2 x_3 - S_f x_3)$$

with an arbitrary C^1 function Φ.

To solve the second PDE we first remember that in the reachability distribution

$$\delta_{22} = -\frac{1}{Y}\delta_{12} \quad \text{and} \quad \delta_{32} = 0$$

and then it's enough to write the PDE as

$$\frac{\partial \lambda}{\partial x_1}\delta_{12}(x) + \frac{\partial \lambda}{\partial x_2}\left(-\frac{1}{Y}\delta_{12}(x)\right) = 0 \tag{6.102}$$

The characteristic equations are written as

$$\dot{x}_1 = \delta_{12}(x)$$
$$\dot{x}_2 = -\frac{1}{Y}\delta_{12}(x)$$
$$\dot{x}_3 = 0$$

It's easy to see that

$$-\frac{1}{Y}\dot{x}_1 - \dot{x}_2 = 0$$

and

$$\dot{x}_3 = 0$$

Therefore the solution of (6.102) is in the form

$$\lambda(x_1, x_2, x_3) = \Phi^*(-\frac{1}{Y}x_1 - x_2, x_3) \tag{6.103}$$

with an arbitrary C^1 function Φ^*. To give a common solution for both (6.101) and (6.102) we propose the function

$$\lambda(x_1, x_2, x_3) = x_3(-\frac{1}{Y}x_1 - x_2 + S_f) = -\frac{1}{Y}x_1 x_3 - (x_2 x_3 - S_f x_3) \quad (6.104)$$

from which we can see that it indeed satisfies both PDEs. With the help of λ we can define the local (and luckily global) coordinates transformation $\Psi : \mathbb{R}^n \to \mathbb{R}^n$

$$\begin{bmatrix} z_1 \\ z_2 \\ z_3 \end{bmatrix} = \Psi(x_1, x_2, x_3) = \begin{bmatrix} x_1 \\ x_2 \\ -\frac{1}{Y}x_1 x_3 - x_2 x_3 + S_f x_3 \end{bmatrix} \quad (6.105)$$

Since

$$x_3 = \frac{z_3}{-\frac{1}{Y}z_1 - z_2 + S_f}$$

and

$$\dot{z}_3 = (-\frac{1}{Y}x_1 x_3 - x_2 x_3 + S_f x_3)'$$

$$= -\frac{1}{Y}(\dot{x}_1 x_3 + x_1 \dot{x}_3) - (\dot{x}_2 x_3 + x_2 \dot{x}_3) + S_f \dot{x}_3$$

$$= -\frac{1}{Y}(\mu(x_2)x_1 x_3 - x_1 + x_1) - (-\frac{1}{Y}\mu(x_2)x_1 x_3 + S_f - x_2 + x_2) + S_f = 0$$

the transformed form of the model (6.77)–(6.79) can be written as

$$\dot{z} = \bar{f}(z) + \bar{g}(z)u \quad (6.106)$$

where

$$\bar{f}(z) = \begin{bmatrix} \frac{\mu_{max} z_2 z_1}{K_1 + z_2 + K_2 z_2^2} \\ -\frac{\mu_{max} z_2 z_1}{(K_1 + z_2 + K_2 z_2^2)Y} \\ 0 \end{bmatrix}, \quad \bar{g}(z) = \begin{bmatrix} -\frac{z_1}{z_3}(-\frac{1}{Y}z_1 - z_2 + S_f) \\ \frac{S_f - z_2}{z_3}(-\frac{1}{Y}z_1 - z_2 + S_f) \\ 0 \end{bmatrix} \quad (6.107)$$

and $z_3 \neq 0$.

6.6.5 Generalizations

The aim of this section is to show the reasons present in the original state-space model which led to the reachability and state transformation above. This analysis enables us to find other models of similar form with the same properties.

Physical Analysis of the Model and the Solutions. The first important thing to observe is that the results of the reachability analysis and that of the coordinates transformation do not depend on the actual form of the function μ in Equation (6.80). The results utilized the following specialities of the original state-space model in Equations (6.77)–(6.79).

1. The constant coefficients in the third state equation, $i.e.$

$$f_3 = 0, \quad g_3 = 1$$

   where f_i and g_i are the i-th entry of the vector functions f and g in
   the state-space model. This property always holds for the overall mass
   balance of fed-batch reactors.
2. The relation between the first and the second state equation, namely

$$f_2 = -\frac{1}{Y}f_1 = C_f f_1$$

   where C_f is a constant. Such a relationship exists if the two related
   state variables, x_1 and x_2, are concentrations of components related by
   a chemical reaction in the form $\frac{1}{Y}S \rightarrow X$ [32].

Further, we may notice that the quantity λ in Equation (6.104) – which is
conserved independently of the input – consists of two parts corresponding
to the substrate mass and cell mass of the system as follows:

$$\lambda(x_1, x_2, x_3) = V(S_f - S) + \frac{1}{Y}V(X_f - X) \tag{6.108}$$

with $X_f = 0$ because the feed does not contain any cells. The above two
terms originate from the (weighted) convective terms in the component mass
conservation balances respectively, that is, such terms which are only caused
by the feed as inflow.

Generalized State-space Models. We can generalize the original model
in Equations (6.77)– (6.79) in two steps if we want to preserve the special
dynamic properties of the model.

1. *General reaction rate function*
   As the results do not depend on the function μ in Equation (6.80), we
   can replace the fermentation reaction by a general chemical reaction of
   the form

$$\frac{1}{Y}S \longrightarrow X$$

   where the reaction rate (source) function is $\mu^*(x_2)x_1$ with μ^* as an un-
   specified possibly nonlinear function.
2. *Non-isotherm case*
   If we further release the assumption that the fermenter is operating under
   isothermal conditions, then we should include the energy conservation
   balance to the original model. Then a four state model is obtained [32]
   in the following input-affine form:

$$\dot{x} = f^*(x) + g^*(x)u \tag{6.109}$$

   where

$$x = \begin{bmatrix} x_1 \\ x_2 \\ x_3 \\ x_4 \end{bmatrix} = \begin{bmatrix} X \\ S \\ T \\ V \end{bmatrix}, \quad u = F \tag{6.110}$$

with T being the temperature in the fermenter. Then the nonlinear functions f^* and g^* are

$$f^*(x) = \begin{bmatrix} \mu^*(x_2, x_3)x_1 \\ c_1\mu^*(x_2, x_3)x_1 \\ c_2\mu^*(x_2, x_3)x_1 \\ 0 \end{bmatrix}, \quad g^*(x) = \begin{bmatrix} -\frac{x_1}{x_4} \\ \frac{S_f - x_2}{x_4} \\ \frac{T_f - x_3}{x_4} \\ 1 \end{bmatrix} \tag{6.111}$$

with $c_1 = -\frac{1}{Y}$ and with the following additional constant parameters:

c_2 reaction enthalpy coefficient $[\text{m}^3\text{K/J}]$
$T_f = 293$ influent temperature $[\text{K}]$

Observe that now the reaction rate function μ^* depends also on the temperature $x_3 = T$, giving rise to the source function $\mu^*(x_2, x_3)x_1$.

Furthermore, the required structural properties 1. and 2. are present in the generalized model. The property 1. now holds for the entries f_4^* and g_4^*, which is the overall mass balance. There are two independent pairs, (f_1^*, f_2^*) (the two component mass balances) and (f_1^*, f_3^*) (a mass and an energy balance) which possess property 2. with different constants.

Analysis of the Generalized Models. In the above four state variable case the final reachability distribution after four steps would be the following:

$$\Delta = \text{span}\{g^*, [f^*, g^*], [f^*, [f^*, g^*]], [g^*, [f^*, g^*]],$$
$$[f^*, [f^*, [f^*, g^*]]], [g^*, [f^*, [f^*, g^*]]],$$
$$[f^*, [g^*, [f^*, g^*]]], [g^*, [g^*, [f^*, g^*]]]\}$$

If we calculate the Lie-products $[f^*, g^*]$, $[f^*, [f^*, g^*]]$ and $[g^*, [f^*, g^*]]$, we find that

$$[f^*, g^*] = \begin{bmatrix} [f^*, g^*]_1 \\ c_1 [f^*, g^*]_1 \\ c_2 [f^*, g^*]_1 \\ 0 \end{bmatrix}, \quad [f^*, [f^*, g^*]] = \begin{bmatrix} [f^*, [f^*, g^*]]_1 \\ c_1 [f^*, [f^*, g^*]]_1 \\ c_2 [f^*, [f^*, g^*]]_1 \\ 0 \end{bmatrix} \tag{6.112}$$

and also

$$[g^*, [f^*, g^*]] = \begin{bmatrix} [g^*, [f^*, g^*]]_1 \\ c_1 [g^*, [f^*, g^*]]_1 \\ c_2 [g^*, [f^*, g^*]]_1 \\ 0 \end{bmatrix} \tag{6.113}$$

Therefore the calculation of the reachability distribution stops here and it turns out that the dimension of the distribution is 2 in this case, too.

To find the decomposed system similar to (6.107), we have to find two independent real-valued functions λ_1 and λ_2 such that

$$\begin{bmatrix} \frac{\partial \lambda_1}{\partial x_1} & \frac{\partial \lambda_1}{\partial x_2} & \frac{\partial \lambda_1}{\partial x_3} & \frac{\partial \lambda_1}{\partial x_4} \\ \frac{\partial \lambda_2}{\partial x_1} & \frac{\partial \lambda_2}{\partial x_2} & \frac{\partial \lambda_2}{\partial x_3} & \frac{\partial \lambda_2}{\partial x_4} \end{bmatrix} \begin{bmatrix} g_1^*(x) & [f^*, g^*]_1(x) \\ g_2^*(x) & c_1 [f^*, g^*]_1(x) \\ g_3^*(x) & c_2 [f^*, g^*]_1(x) \\ 1 & 0 \end{bmatrix} = \begin{bmatrix} 0 & 0 \\ 0 & 0 \end{bmatrix} \tag{6.114}$$

It's easy to check that the two independent functions

$$\lambda_1(x) = x_4(c_1 x_1 - x_2 + S_f) \tag{6.115}$$
$$\lambda_2(x) = x_4(c_2 x_1 - x_3 + T_f) \tag{6.116}$$

satisfy the PDEs in Equation (6.114). Therefore the new coordinate vector z is given by the function $\Psi^* : \mathbb{R}^4 \to \mathbb{R}^4$

$$\begin{bmatrix} z_1 \\ z_2 \\ z_3 \\ z_4 \end{bmatrix} = \Psi^*(x) = \begin{bmatrix} x_1 \\ x_2 \\ x_4(c_1 x_1 - x_2 + S_f) \\ x_4(c_2 x_1 - x_3 + T_f) \end{bmatrix} \tag{6.117}$$

and the system (6.109)–(6.111) in the new coordinates is written as

$$\dot{z} = \bar{f}^*(z) + \bar{g}^*(z)u \tag{6.118}$$

where

$$\bar{f}^*(z) = \begin{bmatrix} \bar{\mu}^*(z_1, z_2, z_3, z_4) z_1 \\ c_1 \bar{\mu}^*(z_1, z_2, z_3, z_4) z_1 \\ 0 \\ 0 \end{bmatrix}, \quad \bar{g}^*(z) = \begin{bmatrix} -\frac{z_1(c_1 z_1 - z_2 + S_f)}{z_3} \\ \frac{(S_f - z_2)(c_1 z_1 - z_2 + S_f)}{z_3} \\ 0 \\ 0 \end{bmatrix} \tag{6.119}$$

and

$$\bar{\mu}^*(z_1, z_2, z_3, z_4) = \mu^* \left(z_2, c_2 z_1 + T_f - \frac{z_4(c_1 z_1 - z_2 + S_f)}{z_3} \right) \tag{6.120}$$

with the condition $z_3 \neq 0$.

6.6.6 Engineering Interpretation

The invariance of λ in Equation (6.104) expresses the fact that the state variables of the fed-batch fermenter model can only move on a smooth hypersurface in the state-space. The shape of this hypersurface obviously depends on the choice of the initial values of the state variables. It means that the initial concentrations and liquid volume (that are set by the control engineer) uniquely determine the set of points in the state-space that are reachable during the process.

Figures 6.1 and 6.2 illustrate the effect of the initial liquid volume on the reachability hypersurface when the concentrations are fixed. It is shown that if the initial volume is too small then the possibilities of controlling the biomass concentration x_1 are dramatically worsening. Similarly, the effect of the initial concentrations can also be easily examined, since λ in Equation (6.104) is a quite simple function of the state variables.

With the help of λ controller design becomes easier. If the desired final point of the fermentation is given (in the state-space) then the initial conditions can be set in such a way that the desired point is reachable.

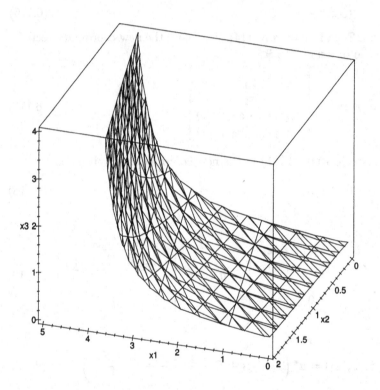

Figure 6.1. The reachability hypersurface of the fed-batch fermenter for initial conditions $x_1(0) = 2\frac{g}{l}$, $x_2(0) = 0.5\frac{g}{l}$, $x_3(0) = 0.5\frac{g}{l}$

6.6.7 Comments on Observability

Due to space limitations we cannot go into detail concerning the observability of fed-batch fermentation processes, we can only briefly describe the most interesting aspect of the relation between reachability and observability. The

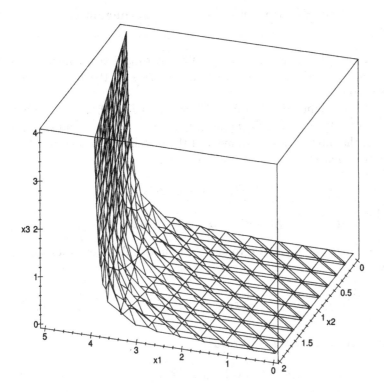

Figure 6.2. The reachability hypersurface of the fed-batch fermenter for initial conditions $x_1(0) = 2\frac{g}{l}$, $x_2(0) = 0.5\frac{g}{l}$, $x_3(0) = 0.1\frac{g}{l}$

rough problem statement of observability is the following: is it possible to determine the values of the state variables of the system if we measure the inputs and the outputs?

Obviously, the observability property of linear and nonlinear systems largely depends on the selection of the output function $y = h(x)$. Let us suppose that the output of the system (6.77)–(6.79) is chosen to be λ in Equation (6.104). Without complicated calculations it is clear, that the system won't be observable because λ is constant in time independently of the input u and therefore it does not provide any information about the internal "movement" of the system. It's value only identifies the reachable hypersurface (manifold).

6.6.8 The Minimal Realization of Fed-batch Fermentation Processes

Using the calculated λ function, it's not difficult to give a minimal state-space realization of fed-batch fermentation processes in the temperature-independent case.

Since the reachability hypersurface defined by λ and shown in Figures 6.1 and 6.2 is two-dimensional, the minimal realization will contain two state variables (i.e. the input-to-state behavior of the system can be described by two differential equations). Since λ is constant in time, it's clear that

$$\lambda(x(t)) = -\frac{1}{Y}x_1(t)x_3(t) - (x_2(t)x_3(t) - S_f x_3(t)) \tag{6.121}$$

$$= -\frac{1}{Y}x_1(0)x_3(0) - (x_2(0)x_3(0) - S_f x_3(0)) = \lambda(x(0))$$

Therefore we can express, e.g. the volume x_3 from the above equation in the following way:

$$x_3 = \frac{\lambda(x(0))}{-\frac{1}{Y}x_1 + S_f - x_2}, \quad -\frac{1}{Y}x_1 + S_f - x_2 \neq 0 \tag{6.122}$$

and the minimal state-space model reads

$$\dot{x} = f_{min}(x) + g_{min}(x)u$$

where

$$x = \begin{bmatrix} x_1 \\ x_2 \end{bmatrix}, \quad f_{min}(x) = \begin{bmatrix} \frac{\mu_{max}x_2 x_1}{K_1 + x_2 + K_2 x_2^2} \\ -\frac{\mu_{max}x_2 x_1}{(K_1 + x_2 + K_2 x_2^2)Y} \end{bmatrix}$$

$$g_{min}(x) = \begin{bmatrix} \frac{\frac{1}{Y}x_1^2 + x_1(x_2 - S_f)}{\lambda(x(0))} \\ \frac{(-\frac{1}{Y}x_1 + S_f - x_2)(S_f - x_2)}{\lambda(x(0))} \end{bmatrix} \tag{6.123}$$

It's well-known from system theory that state-space realizations are not unique and it's easy to see that instead of x_3 any one of the other two state variables could be expressed from Equation (6.121). Therefore one can select those two state variables that are important from a certain point of view (e.g. a control problem) and express the third one from Equation (6.121).

It's also important to remark that the model (6.123) has a special structure, since it contains the initial values of the original model (6.77)–(6.79) in the input vector field g_{min} (but luckily not in the vector field f_{min}).

6.7 Further Reading

The controllability and observability of linear systems is covered in several good textbooks, e.g. in [39], [20] or [13], to mention just a few. Although [7] is mainly devoted to linear systems, it discusses important connections to nonlinear systems.

Additional results about the controllability of nonlinear systems can be found in [70] and [73]. The observability and the problem of identifying the initial state in a single-input nonlinear system is studied in [72]. The reachability and observability of bilinear systems are discussed in [14].

6.8 Summary

Based on the controllability and observability notions and analysis tools developed for LTI systems, the basic concepts and algorithms for analyzing state controllability and observability of input-affine nonlinear systems are described in this section.

The nonlinear notions and methods are illustrated using case studies of process systems of practical importance: heat exchangers and fermenters of various type. Their controllability and observability analysis is based upon the special nonlinear structure of process systems and on their input-affine nonlinear state-space model developed in Chapter 4.

6.9 Questions and Application Exercises

Exercise 6.9.1. Compare the notions and tools related to linear and nonlinear controllability analysis. Comment on the obtained analysis results and on the computational resources needed.

Exercise 6.9.2. Compare the notions and tools related to linear and nonlinear observability analysis. Comment on the obtained analysis results and on the computational resources needed.

Exercise 6.9.3. Perform linear controllability and observability analysis of the nonlinear heat exchanger cell described in Section 4.4.4 using its linearized state-space model developed in Exercise 4.8.10.

Exercise 6.9.4. Consider the following two-dimensional nonlinear system model in its input-affine state-space form:

$$\frac{dx_1}{dt} = x_1^{1/2} x_2^{3/2} - x_1 x_2^3 + x_1^{1/2} u_1$$

$$\frac{dx_2}{dt} = 3x_1 x_2^{1/2} - 2x_1^2 x_2$$

$$y = x_1^3 x_2^{2/3} - x_1^{1/2} x_2^2$$

Compute the controllability distribution Δ_c for this system and determine if it is controllable or not.

Exercise 6.9.5. Consider the following two-dimensional nonlinear system model in its input-affine state-space form:

$$\frac{dx_1}{dt} = -\frac{x_1}{x_2^2 + 2x_1 + 3} - x_1 u_1$$

$$\frac{dx_2}{dt} = 4\frac{x_1}{x_2^2 + 2x_1 + 3} - x_2 u_1 - \frac{x_1}{x_2} u_2$$

$$y = \frac{x_1}{x_2}$$

Compute the controllability distribution Δ_c for this system and determine its rank.

Exercise 6.9.6. The following two-dimensional nonlinear system model is given:

$$\frac{dx_1}{dt} = -x_1 e^{-x_2} - x_1 u_1$$

$$\frac{dx_2}{dt} = 2x_1 e^{-x_2} - x_2 u_1 + x_2 u_2$$

$$y = e^{x_1}$$

Determine the controllability of the system using its controllability distribution Δ_c.

7. Stability and The Lyapunov Method

Stability is a basic dynamic system property. It characterizes the behavior of a system if that is subject to disturbances. In general everyday terms, a system is regarded to be "stable" if it is not disturbed too much by the external disturbances, but rather it dampens their effect.

There are various but related notions in systems and control theory to characterize the stability of a system. Internal or asymptotic stability focuses on the effect of the disturbance on the state of the system in a long time horizon, while external or bounded-input–bounded-output (BIBO) stability deals only with the effect seen in the system outputs.

The Lyapunov method for checking asymptotic stability is the technique which is applicable for nonlinear systems. Therefore we summarize the notions of stability and the Lyapunov method here in this chapter together with checking stability for both linear and nonlinear systems.

The material is arranged in the following sections:

- *Stability notions*
  Here we describe both the bounded-input–bounded-output (BIBO) and the asymptotic stability case. First, the general notions are given, then we illustrate them in the LTI case by stating necessary and sufficient conditions.
- *Local stability of nonlinear systems*
  Thereafter, the local stability of the nonlinear case is discussed together with the description of local linearization techniques.
- *Lyapunov's theorem and Lyapunov functions*
  We start with the notion of the Lyapunov function as a generalized energy function. Thereafter Lyapunov's theorem on asymptotic stability is stated. As an illustration the Lyapunov conditions for the LTI case are also given.
- *Stability of process systems*
  Besides the analytical techniques for analyzing stability, the notion of structural stability and an algebraic method for its analysis is introduced.
- *Examples and a case study*
  The stability analysis techniques are illustrated using simple well-known process examples (heat exchanger cells, mass convection network and a binary distillation tray) and on a simple continuous fermenter.

7.1 Stability Notions

There are two different but related notions of stability:

1. *Bounded-input–bounded output (BIBO) stability*, which describes the behavior of the system if it is subject to bounded but permanent disturbances.
2. *Asymptotic stability* when the disturbance acts as an impulse (over an infinitesimally short time interval) and then the system behavior is analyzed when time goes to infinity.

7.1.1 External or BIBO (Bounded-input–bounded-output) Stability

In order to define BIBO stability, we need to recall the notions of vector and signal norms which are summarized in Appendix A in Section A.1.

Definition 7.1.1 (BIBO stability)
A system is external or BIBO-stable if for any bounded input it responds with a bounded output

$$\{||u(t)|| \leq M_1 < \infty \mid 0 \leq t \leq \infty\} \Rightarrow \{||y(t)|| \leq M_2 < \infty \mid 0 \leq t \leq \infty\} \tag{7.1}$$

where $||.||$ is a vector norm.

Observe that the definition above says nothing about the states but uses the concept of input–output representation of the system **S**. If we denote the signal norm of the input by $||u||$ and that of the output by $||u||$ then the defining equation (7.1) can be interpreted as a condition for the system operator **S**:

$$||\mathbf{S}|| < \infty \tag{7.2}$$

where $||\mathbf{S}||$ is the induced operator norm by the signal norms for the input and output signals.

7.1.2 BIBO Stability Conditions for LTI Systems

Recall from Chapter 5 that the impulse-response function h or the transfer function H is used to describe the abstract system operator **S** of LTI systems. Therefore it is natural to look for conditions of their BIBO stability in terms of the properties of their impulse-response or transfer function.

The theorem below gives a simple example of such results for the case of single-input–single-output LTI systems.

Theorem 7.1.1 (BIBO stability of SISO LTI system). *A single-input–single-output (SISO) LTI system is (externally or) BIBO-stable if and only if*

$$\int_0^\infty |h(t)|dt \leq M < \infty \tag{7.3}$$

where M is a constant and $h(t)$ is the impulse-response function of the system.

7.1.3 L_2-gain of Linear and Nonlinear Systems

The concept of L_2-gain (see Subsection 2.2.2) of an LTI system plays a central role in the investigation of stability and performance of control systems. In this section we extend and refine this concept to a class of nonlinear systems, too.

In order to define the L_2-gain of a linear system we go back to the original definition of BIBO stability given by Equation (7.1) and develop a special case of it.

Definition 7.1.2 (L_2-gain for linear systems)
A linear system with input u and output y has L_2-gain less than or equal to γ, if

$$\int_0^\infty \|y\|^2 dt \leq \gamma^2 \int_0^\infty \|u\|^2 dt \tag{7.4}$$

Observe that the integrals in the above inequality are the 2-norms of the signals u and y, that is

$$\int_0^\infty \|y\|^2 dt = \|y\|_2^2$$

We assume that the squared 2-norm $\int_0^\infty \|u\|^2$ is finite in Equation (7.4). In this case the BIBO stability of the system implies that there exists a finite γ^2 such that it holds.

> It is known that finite L_2-gain implies BIBO stability of a linear system.

In the theory of nonlinear systems, however, the notion of L_2-gain has a slight difference when compared to the defining Equation (7.4).

Consider a nonlinear system given by an input-affine nonlinear state-space model:

$$\begin{aligned} \dot{x} &= f(x) + g(x)u & u \in \mathbb{R}^m, \; f(0) = 0 \\ y &= h(x) & y \in \mathbb{R}^p, \; h(0) = 0 \end{aligned} \tag{7.5}$$

where $x = (x_1, \ldots, x_n)$ are locale coordinates for a e.g. C^k (k times continuously differentiable functions, $k \geq 1$) state-space manifold $\mathcal{M}$.

The L_2-gain concept can be carried over to the system described in Equation (7.5) in the following form:

Definition 7.1.3 (L_2-gain for nonlinear systems)
The nonlinear system (7.5) has L_2-gain less than or equal to γ, if for all
$x \in \mathcal{M}$, $\exists K(x)$, $0 \le K(x) < \infty$, $K(0) = 0$, *such that for every $T > 0$:*

$$\int_0^T \|y\|^2 dt \le \gamma^2 \int_0^T \|u\|^2 dt + K(x(T)) \tag{7.6}$$

Observe that the function $K(.)$ gives the "remaining part" of the integrals in Equation (7.4) from T up to ∞ and therefore it should be taken at the point $x(T)$. Furthermore, it is clear that the function $K(.)$ is not unique.

The function $K(x)$ in the above definition gives rise to the notion of the *available storage*, which leads to the *storage function* of the system (see later in Chapter 8).

7.1.4 The Small-gain Theorem

For closed-loop systems the celebrated small-gain theorem can be used to investigate BIBO stability. The small-gain theorem is based on the operator description of a composite system consisting of two interconnected subsystems. The notations and results presented in this subsection are based on [80], where the complete proof of the theorem can be found.

For the general statement of the small-gain theorem it is necessary to define the notion of L_q-stability and finite L_q-gain of relations instead of maps.

Definition 7.1.4 (L_q-stability and finite L_q-gain of relations)
- $R \subset L_{qe}(U) \times L_{qe}(Y)$ *is L_q-stable if for all*
  $(u, y) \in R$, $u \in L_q(U) \Rightarrow y \in L_q(Y)$,
- $R \subset L_{qe}(U) \times L_{qe}(Y)$ *has finite L_q-gain if $\exists \gamma_q, b_q \in \mathbb{R}$*
  such that for all $T \ge 0$
  $(u, y) \in R$, $u \in L_{qe}(U) \Rightarrow \|y_T\|_q \le \gamma_q \|u_T\|_q + b_q$.

General Feedback Configuration. The general feedback configuration consists of two interconnected subsystems with their system operators G_1 and G_2, and is shown in Figure 7.1.

The signals appearing in the closed-loop system $\Sigma^f_{G_1, G_2}$ are as follows:

$$u_1(t), y_2(t), e_1(t) \in E_1, \quad \forall t \ge 0 \tag{7.7}$$
$$u_2(t), y_1(t), e_2(t) \in E_2, \quad \forall t \ge 0 \tag{7.8}$$

where $\dim(E_1) = m$ and $\dim(E_2) = n$.
The subsystems forming $\Sigma^f_{G_1, G_2}$ are

$$G_1 : L_{qe}(E_1) \mapsto L_{qe}(E_2), \quad G_2 : L_{qe}(E_2) \mapsto L_{qe}(E_1) \tag{7.9}$$

Furthermore, we assume that $e_1 \in L_{qe}(E_1)$ and $e_2 \in L_{qe}(E_2)$.
It's clear from Figure 7.1 that

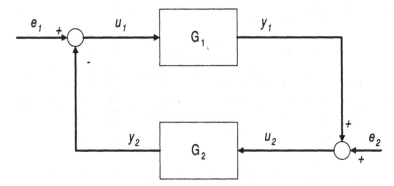

Figure 7.1. General feedback configuration

$$u_1 = e_1 - y_2, \; u_2 = e_2 + y_1$$
$$y_1 = G(u_1), \quad y_2 = G_2(u_2) \tag{7.10}$$

Let us use the following notations

$$u = \begin{bmatrix} u_1 \\ u_2 \end{bmatrix}, \quad y = \begin{bmatrix} y_1 \\ y_2 \end{bmatrix}, \quad e = \begin{bmatrix} e_1 \\ e_2 \end{bmatrix} \tag{7.11}$$

Then the feedback connection of the two subsystems is described as

$$y = G(u) \text{ and } u = e - Fy \tag{7.12}$$

where

$$G = \begin{bmatrix} G_1 & 0 \\ 0 & G_2 \end{bmatrix} \text{ (operator matrix)}, \quad F = \begin{bmatrix} 0 & I_{m_1} \\ -I_{m_2} & 0 \end{bmatrix} \text{ (real matrix)} \tag{7.13}$$

Then the overall input and output signals are expressed as

$$y = G(e - Fy) \tag{7.14}$$
$$u = e - FG(u) \tag{7.15}$$

The closed-loop system defines two relations:

$$R_{eu} = \{(e, u) \in L_{qe}(E_1 \times E_2) \times L_{qe}(E_1 \times E_2) \mid u + FG(u) = e\} \tag{7.16}$$
$$R_{ey} = \{(e, y) \in L_{qe}(E_1 \times E_2) \times L_{qe}(E_2 \times E_1) \mid y = G(e - Fy)\} \tag{7.17}$$

The most important properties of the closed-loop system are the following:

- R_{eu} is L_q-stable $\Leftrightarrow$ R_{ey} is L_q-stable,
- R_{eu} has finite L_q-gain $\Leftrightarrow$ R_{ey} has finite L_q-gain,
- let G_1 and G_2 be causal I–O mappings corresponding to causal systems (see Section 2.2.1). Then $\Sigma^f_{G_1, G_2}$ is also causal in the following sense:
  1. u_T depends only on e_T for all $(e, u) \in R_{eu}$ and $T \geq 0$.

2. y_T depends only on e_T for all $(e, y) \in R_{ey}$ and $T \geq 0$.

Now we can state the small-gain theorem in its most general form as follows:

Theorem 7.1.2 (Small-gain theorem). *Consider the closed-loop system* Σ_{G_1,G_2}^f *and let* $q \in \{1, 2, \ldots, \infty\}$. *Suppose that* G_1 *and* G_2 *have* L_q-gains $\gamma_q(G_1)$ *and* $\gamma_q(G_2)$ *respectively. The closed-loop system has finite* L_q-gain *if*

$$\gamma_q(G_1) \cdot \gamma_q(G_2) < 1$$

7.1.5 Asymptotic or Internal Stability of Nonlinear Systems

Asymptotic stability corresponds to a situation where the effect of an impulse-type disturbance moves out the state of the system from its non-perturbed (nominal) value over an infinitesimally short time interval and then the perturbed state evolution is compared to the nominal one. As we are interested in the time evolution of the state variable(s), we need to have a state-space model of the system for analyzing asymptotic stability. Also, the focus is on the behavior of the state variables, these being internal signals of the system: it explains why asymptotic stability is also called *internal stability*.

Because of the above situation, we consider identically zero inputs, *i.e.* $u(t) \equiv 0$ in the state-space model of the system when analyzing asymptotic stability.

Definition 7.1.5 (Truncated state equation)
A state equation with identically zero inputs is called a truncated state equation.

In order to give a rigorous mathematical definition of asymptotic stability in the general nonlinear case, let us have a truncated nonlinear state equation (with $u(t) \equiv 0$) as follows:

$$\frac{dx}{dt} = f(x, t) \tag{7.18}$$

and let it have two solutions:

- $x^0(t)$ for $x^0(t_0)$ as the ordinary solution, and
- $x(t)$ for $x(t_0)$ which is a "perturbed solution".

Definition 7.1.6 (Stability of a solution)
The solution $x^0(t)$ *of Equation (7.18) is stable if for any given* $\epsilon > 0$ *there exists a* $\delta(\epsilon, t_0)$ *such that all solutions with* $||x(t_0) - x^0(t_0)|| < \delta$ *fulfill* $||x(t) - x^0(t)|| < \epsilon$ *for all* $t \geq t_0$ *where* $||.||$ *is a suitable vector norm.*

The above stability notion is used to state the definition of asymptotic stability.

Definition 7.1.7 (Asymptotic stability, weak)
The solution $x^0(t)$ of Equation (7.18) is asymptotically stable if it is stable and $||x(t) - x^0(t)|| \to 0$ when $t \to \infty$ provided that $||x(t_0) - x^0(t_0)||$ is small enough.

The above definitions have a very special property as compared to the notion of BIBO stability.

> *Asymptotic stability is a property of a solution for nonlinear systems.*

Asymptotic Stability of Input-affine Nonlinear Systems. In the case of input-affine nonlinear systems, we assume that $f(x,t) = f(x)$ and $f(0) = 0$. Then Equation (7.18) has a stationary solution $x^0(t) \equiv 0$ for $x^0(t_0) = 0$. In this case, the notion of *asymptotic stability* is used in a more strict sense as follows:

Definition 7.1.8 (Asymptotic stability, strong)
Equation (7.18) is asymptotically stable if $||x(t)|| \to 0$ when $t \to \infty$ provided that $||x(t_0)||$ is small enough.

7.1.6 Asymptotic Stability of LTI Systems

In order to investigate the similarities and differences of asymptotic stability for nonlinear and linear systems, we start with the usual definition of asymptotic stability special to LTI systems.

Definition 7.1.9 (Asymptotic stability of LTI systems)
An LTI system with realization (A, B, C) is (internally) stable if the solution $x(t)$ of the equation

$$\dot{x}(t) = Ax(t), \quad x(t_0) = x_0 \neq 0, \quad t > t_0 \tag{7.19}$$

fulfills

$$\lim_{t \to \infty} x(t) = 0$$

The condition above is similar to asymptotic stability of linear ordinary differential equations. Moreover, this notion is exactly the same as the strong asymptotic stability but there is no need for $||x(t_0)||$ to be "small enough". The reason for this difference is explained later.

In order to develop necessary and sufficient conditions for an LTI system to be asymptotically stable, the following definition will be useful:

Definition 7.1.10 (Stability matrix)
A square matrix $A \in \mathbb{R}^{n \times n}$ is said to be a stability matrix if all of its eigenvalues $\lambda_i(A)$ have strictly negative real parts, i.e.

$$Re\{\lambda_i(A)\} < 0, \forall i$$

The following proposition shows the effect of state transformations on the stability matrix property.

Proposition 7.1.1. *The eigenvalues of a square $A \in \mathbb{R}^{n \times n}$ matrix remain unchanged after a similarity transformation on A, i.e. with*

$$\bar{A} = T A T^{-1}$$

we have the same eigenvalues for both A and $\bar{A}$.

Now we are in a position to state the necessary and sufficient condition for an LTI system to be asymptotically stable.

Theorem 7.1.3 (Asymptotic stability of LTI systems). *An LTI system is asymptotically (internally) stable if and only if its state matrix A is a stability matrix.*

The above proposition and theorem together show that the stability property of a system remains unchanged under similarity transformation, that is, asymptotic stability is *realization-independent*.

It is important to emphasize that stability is a system property:

- *This is true by construction for BIBO stability.*
- *The above Proposition 7.1.1 and Theorem 7.1.3 proves it for the asymptotic stability of LTI systems.*

7.1.7 Relationship Between Asymptotic and BIBO Stability

Generally speaking, asymptotic stability is a "stronger" notion than BIBO stability. This is because it is intuitively clear that a bounded state signal produces a bounded output signal through a smooth and bounded output function $y = h(x)$. Therefore we expect that asymptotic stability implies BIBO stability under relatively mild conditions.

The following theorem shows the relationship between BIBO and asymptotic stability in the case of LTI systems.

Theorem 7.1.4 (Asymptotic and BIBO stability of LTI systems). *Asymptotic stability implies BIBO stability for LTI systems but the reverse statement does not hold.*

The following two simple examples show that a BIBO stable system is indeed not necessarily asymptotically stable.

Example 7.1.1 (Asymptotic and BIBO stability 1)
A BIBO-stable but not asymptotically stable LTI system

Consider a LTI system with the state-space representation

$$\dot{x} = \begin{bmatrix} -2 & 0 \\ 0 & 3 \end{bmatrix} x + \begin{bmatrix} 2 \\ 1 \end{bmatrix} u \tag{7.20}$$

$$y = \begin{bmatrix} 1 & 0 \end{bmatrix} \tag{7.21}$$

The system above is not asymptotically stable since one of the eigenvalues of the state matrix is positive, but it is BIBO-stable because the "unstable state" does not appear in the output.
The above system is called *non-detectable*, since the unstable state is clearly non-observable.

Example 7.1.2 (Asymptotic and BIBO stability 2)
Another BIBO-stable but not asymptotically stable LTI system

Consider an LTI system with the truncated state equation

$$\dot{x} = \begin{bmatrix} 0 & -2 \\ 2 & 0 \end{bmatrix} x \tag{7.22}$$

The eigenvalue analysis shows that

$$\lambda_{1,2} = \pm i$$

that is, the poles of the system are on the stability boundary. This system is a pure oscillator.

7.2 Local Stability of Nonlinear Systems

One of the most widespread approach for analyzing local asymptotic stability of a nonlinear system in the neighborhood of a steady-state operating point is to linearize the system model around the operating point and then perform linear stability analysis. This approach is the subject of this section, where we discuss how to linearize nonlinear state-space models, how to relate local and global analysis results and how local stability depends on the system parameters.

7.2.1 Local Linearization of Nonlinear State-space Models

In order to linearize a state or output equation in a nonlinear state-space model, we first need to find a steady-state operating or reference point to linearize around and then proceed with the linearization.

The Reference Point. The reference point (u_0, x_0, y_0) for the linearization is found by specifying a reference input u_0 for a nonlinear state-space model in its general form:

$$\dot{x} = f(x, u), \quad y = h(x, u) \tag{7.23}$$

and computing the reference value of the state and output signals from the above equations by specifying $\dot{x} = 0$ to have a steady-state operating point:

$$\widetilde{f}(x_0, u_0) = 0, \quad y_0 = \widetilde{h}(x_0, u_0) \tag{7.24}$$

In the special case of input-affine state-space models, we have:

$$f(x_0) + \sum_{i=1}^{m} g_i(x_0) u_{i0} = 0, \quad y_0 = h(x_0) \tag{7.25}$$

from the state-space model

$$\dot{x}(t) = f(x(t)) + \sum_{i=1}^{m} g_i(x(t)) u_i(t)$$
$$y(t) = h(x(t)) \tag{7.26}$$

Thereafter, *centered variables* are introduced for all variables to have

$$\widetilde{x} = u - u_0, \quad \widetilde{x} = x - x_0, \quad \widetilde{y} = y - y_0 \tag{7.27}$$

The Principle of Linearization. The linearization is based on a Taylor-series expansion of a smooth nonlinear function f around a steady-state reference point x_0.

In the case of a univariate (*i.e.* when x is scalar) scalar-valued function $y = f(x)$, the following Taylor series expansion is obtained:

$$f(x) = f(x_0) + \frac{df}{dx}\bigg|_{x_0} (x - x_0) + \frac{d^2 f}{dx^2}\bigg|_{x_0} \frac{(x - x_0)^2}{2!} + \ldots \tag{7.28}$$

where x_0 is the reference point and $\frac{d^i f}{dx^i}$ denotes the i-th partial derivative of the function f with respect to x.

The linear approximation is then obtained by neglecting the higher-order terms, which gives:

$$f(x) \simeq f(x_0) + \left.\frac{df}{dx}\right|_{x_0} (x - x_0) \tag{7.29}$$

This way, we obtain a linear approximation of the original nonlinear equation $y = f(x)$ in the form:

$$y = f(x_0) + \left.\frac{df}{dx}\right|_{x_0} (x - x_0) \tag{7.30}$$

The above equation can be simplified to obtain

$$(\widetilde{y} + y_0) = f(x_0, t) + \left.\frac{df}{dx}\right|_{x_0} (x - x_0) \tag{7.31}$$

because

$$y_0 = f(x_0) \quad \text{and} \quad \widetilde{y} = y - y_0 \tag{7.32}$$

Finally we obtain

$$\widetilde{y} = \left.\frac{df}{dx}\right|_{x_0} \widetilde{x} \tag{7.33}$$

which is the final linearized form expressed in terms of the centered variable $\widetilde{x}$.

It is easy to extend the above formulae (7.33) for the case of a multivariate vector-valued function $y = f(x_1, \ldots, x_n, t)$ to obtain

$$\widetilde{y} = \left.J^{f,x}\right|_{x_0} \cdot \widetilde{x} \tag{7.34}$$

where $J^{f,x}$ is the so-called Jacobian matrix of the function f containing the partial derivatives of the function with respect to the variable x evaluated at the steady-state reference point

$$\left.J_{ji}^{f,x}\right|_0 = \left.\frac{\partial f_j}{\partial x_i}\right|_{x_0} \tag{7.35}$$

The Linearized Form of the State-space Model. Now we can apply the principle of linearization to the case of nonlinear state-space models. To do this, the nonlinear functions $\widetilde{f}$ and $\widetilde{h}$ in the general state-space model in Equation (7.23) or the nonlinear functions f, g and h in Equation (7.26) can be linearized separately around the same steady-state reference point (u_0, x_0, y_0).

The linearized model can then be expressed in terms of centered variables in an LTI form:

$$\begin{aligned} \dot{\widetilde{x}} &= \widetilde{A}\widetilde{x} + \widetilde{B}\widetilde{u} \\ \widetilde{y} &= \widetilde{C}\widetilde{x} + \widetilde{D}\widetilde{u} \end{aligned} \tag{7.36}$$

where in the general nonlinear state-space model case the matrices can be computed as

$$\tilde{A} = J^{\tilde{f},x}\Big|_0, \quad \tilde{B} = J^{\tilde{f},u}\Big|_0, \quad \tilde{C} = J^{\tilde{h},x}\Big|_0, \quad \tilde{D} = J^{\tilde{h},u}\Big|_0 \tag{7.37}$$

while in the input-affine case they are in the form:

$$\tilde{A} = J^{f,x}\Big|_0 + J^{g,x}\Big|_0\, u_0, \quad \tilde{B} = g(x_0), \quad \tilde{C} = J^{h,x}\Big|_0, \quad \tilde{D} = 0 \tag{7.38}$$

7.2.2 Relationship Between Local and Global Stability of Nonlinear Systems

The relationship between local and global stability is far from being trivial. It is intuitively clear that global stability is a stronger notion, that is, a locally stable steady-state point may not be globally stable. This is the case when more than one steady-state point exists for a nonlinear system.

There is a related but important question concerning local stability of nonlinear systems: *how and when can we draw conclusions on the local stability of a steady-state point of a nonlinear system from the eigenvalues of its locally linearized state matrix?* The following theorems provide answers to this question.

Theorem 7.2.1 (Theorem for proving local asymptotic stability). *Suppose that each eigenvalue of the matrix $\tilde{A}$ in Equation (7.36) has a negative real part. Then x_0 is a locally asymptotically stable equilibrium of the original system (7.26).*

The following theorem shows how one can relate the eigenvalues of a locally linearized state matrix to the unstable nature of the equilibrium point.

Theorem 7.2.2 (Theorem for proving instability). *Suppose that the matrix $\tilde{A}$ in (7.36) has at least one eigenvalue with a positive real part. Then x_0 is an unstable equilibrium of the original system (7.26).*

The following simple examples illustrate the application of the above results to simple systems.

Example 7.2.1 (Global and local asymptotic stability 1)

We show that Theorem 7.2.1 is only true in one direction. For this purpose, consider the one-dimensional autonomous system

---

$$\dot{x} = -x^3 \tag{7.39}$$

The only steady-state equilibrium of this system is $x_0 = 0$.
The Jacobian matrix is now a constant $J^{f,x} = -3x^2$ evaluated at $x_0 = 0$, which is again 0.
It's easy to check that for any initial condition $x(0)$

$$x(t) = \frac{x(0)}{\sqrt{2tx(0)^2 + 1}} \tag{7.40}$$

From this, it follows that

$$|x(t)| \le |x(0)|, \quad \forall t \ge 0 \tag{7.41}$$

and obviously

$$\lim_{t \to \infty} x(t) = 0 \tag{7.42}$$

Therefore 0 is a globally asymptotically stable equilibrium of the system (7.39).

Example 7.2.2 (Global and local asymptotic stability 2)

In this simple example we show that only stability of the linearized system is not enough even for the stability of the original nonlinear system. For this, consider the one-dimensional autonomous system

$$\dot{x} = -x^2 \tag{7.43}$$

It's easy to see that the linearized system (around the equilibrium state $x_0 = 0$) reads

$$\dot{\tilde{x}} = 0 \tag{7.44}$$

which is stable (but not asymptotically stable). It can be checked that the solution of Equation (7.43) for the initial state $x(0)$ is given by

$$x(t) = \frac{x(0)}{tx(0) + 1}, \quad \forall t \geq 0 \tag{7.45}$$

This means that 0 is an unstable equilibrium state of (7.43) since the solution is not even defined for all $t \geq 0$ if $x(0) < 0$.

Example 7.2.3 (Asymptotic stability, nonlinear case)

Consider the following two-dimensional model (which is the model of a mathematical pendulum with rod length l and gravity constant g)

$$\dot{x}_1 = x_2 \tag{7.46}$$

$$\dot{x}_2 = -\frac{g}{l}\sin(x_1) \tag{7.47}$$

It's easy to see that this system has exactly two equilibrium points, namely $x_{01} = [0 \ 0]^T$ and $x_{02} = [\pi \ 0]^T$. The system matrix of the linear approximation in x_{02} is given by

$$\tilde{A} = \begin{bmatrix} 0 & 1 \\ \frac{g}{l} & 0 \end{bmatrix} \tag{7.48}$$

which has an eigenvalue with a positive real part, therefore x_{02} is an unstable equilibrium of the system according to Theorem 7.2.2. We can't say anything about the stability of x_{01} based on the linear approximation. However, it's not difficult to prove with other methods that x_{01} is a locally stable but not asymptotically stable equilibrium state.

7.2.3 Dependence of Local Stability on System Parameters: Bifurcation Analysis

Roughly speaking, bifurcations represent the sudden appearance of a qualitatively different solution for a nonlinear system as some parameter varies. We speak about a bifurcation when the topological structure of the phase portrait of a dynamical system changes when a parameter value is slightly changed.

Definition 7.2.1 (Topological equivalence)
Let $f : M \mapsto \mathbb{R}^n$ and $g : N \mapsto \mathbb{R}^n$ be class C^r mappings where $M, N \subset \mathbb{R}^n$ are open sets, and consider the autonomous systems

$$\dot{x} = f(x) \tag{7.49}$$
$$\dot{x} = g(x) \tag{7.50}$$

with flows existing for all $t \in \mathbb{R}$. The two differential equations (7.49) and (7.50) are topologically equivalent if there exists a homeomorphism $h : M \mapsto N$ that transforms the trajectories of Equation (7.49) into the trajectories of Equation (7.50) keeping the direction.

If Equation (7.49) and Equation (7.50) are topologically equivalent, then h maps equilibrium points onto equilibrium points and periodic trajectories onto periodic trajectories (by possibly changing the period).

In order to define bifurcation values, consider a parameter-dependent unforced nonlinear system

$$\dot{x} = f(x, \epsilon), \quad x \in M, \quad \epsilon \in V \tag{7.51}$$

where $M \subset \mathbb{R}^n$ and $V \subset \mathbb{R}^l$ are open sets and $f : M \times V \mapsto \mathbb{R}^n$ is a class C^r function. Notice that Equation (7.51) can be also be interpreted as a parameter-independent system $\dot{x} = f(x, \epsilon)$, $\dot{\epsilon} = 0$. We assume that (7.51) has a solution on the entire set $\mathbb{R}$, and $f(0,0) = 0$.

Definition 7.2.2 (Bifurcation value)
*The parameter $\epsilon = 0 \in V$ is called a bifurcation value if in any neighborhood of 0 there exist $\epsilon \in V$ parameter values such that the two autonomous systems $\dot{x} = f(x, 0)$ and $\dot{x} = f(x, \epsilon)$ are **not** topologically equivalent.*

Besides characterizing bifurcation points, Theorem 7.2.3 below can also be used for investigating local stability based on the eigenvalues of the linearized state matrix with multiple zero eigenvalues.

Let $\lambda_1, \ldots \lambda_s$ be those eigenvalues of the linearized system matrix $\widetilde{A}$ that have zero real parts, where

$$\widetilde{A} = D_x f(0,0) = \left[\frac{\partial f_i}{\partial x_j}(0,0) \right]_{i,j=1}^n \tag{7.52}$$

Furthermore, let $\widetilde{A}$ have exactly m eigenvalues with negative real parts and $k = n - s - m$ eigenvalues with positive real parts.

Theorem 7.2.3 (Center manifold theorem). *In a neighborhood of 0 with ϵ of a sufficiently small norm, the system (7.51) is topologically equivalent with the following system:*

$$\dot{x} = F(x, \epsilon) := Hx + g(x, \epsilon)$$
$$\dot{y} = -y \tag{7.53}$$
$$\dot{z} = z \tag{7.54}$$

where $x(t) \in \mathbb{R}^s$, $y(t) \in \mathbb{R}^m$, $z(t) \in \mathbb{R}^k$ and H is a quadratic matrix of size s with eigenvalues $\lambda_1 \ldots \lambda_s$. Furthermore, g is a class C^r function for which $g(0,0) = 0$ and $D_x g(0,0) = 0$.

From the construction of Equation (7.53), it follows that in a neighborhood of 0, the bifurcations of Equation (7.51) can be described using only the equation

$$\dot{x} = F(x, \epsilon) \tag{7.55}$$

Equation (7.55) is called the *reduced differential equation* of Equation (7.53) on the *local center manifold* $M_{loc}^c = \{(x, y, z)|y = 0, z = 0\}$.

Often, Equation (7.55) can be transformed into a simpler (*e.g.* polynomial) form using a nonlinear parameter-dependent coordinate transformation which does not alter the topological structure of the phase portrait in a neighborhood of the investigated equilibrium point. This transformed form is called a *normal form*. The normal form is not unique, different normal forms can describe the same bifurcation equivalently.

Definition 7.2.3 (Fold bifurcation)
Let $f : \mathbb{R} \times \mathbb{R}$ be a one-parameter C^2 class map satisfying

$$f(0, 0) = 0 \tag{7.56}$$

$$\left[\frac{\partial f}{\partial x}\right]_{\epsilon=0, x=0} = 0 \tag{7.57}$$

$$\left[\frac{\partial^2 f}{\partial x^2}\right]_{\epsilon=0, x=0} > 0 \tag{7.58}$$

$$\left[\frac{\partial f}{\partial \epsilon}\right]_{\epsilon=0, x=0} > 0 \tag{7.59}$$

there then exist intervals $(\epsilon_1, 0)$, $(0, \epsilon_2)$ and $\mu > 0$ such that:

- *if $\epsilon \in (\epsilon_1, 0)$, then $f(\cdot, \epsilon)$ has two equilibrium points in $(-\mu, \mu)$ with the positive one being unstable and the negative one stable, and*
- *if $\epsilon \in (0, \epsilon_2)$, then $f(\cdot, \epsilon)$ has no equilibrium points in $(-\mu, \mu)$.*

This type of bifurcation is known as a fold bifurcation (also called a saddle-node bifurcation or tangent bifurcation).

7.3 Lyapunov Function, Lyapunov Theorem

The Lyapunov theorem and the Lyapunov function, the existence of which is the condition of the theorem, play a central role in analyzing asymptotic stability of nonlinear systems. This is the most widespread and almost the only generally applicable technique for this case.

7.3.1 Lyapunov Function and Lyapunov Criterion

The concept of a Lyapunov function originates from theoretical mechanics. Here we see that in stable conservative systems "energy" is a positive definite scalar function which should decrease with time. Using this analogy, we can define a *generalized energy* as a *Lyapunov* function to analyze stability for any nonlinear system.

Definition 7.3.1 (Lyapunov function)

A generalized energy or Lyapunov function $V[x]$ to an autonomous system with a truncated state equation $\dot{x} = f(x)$ is a scaler-valued function with the following properties:

1. *scalar function*

$$V : \mathbb{R}^n \rightarrow \mathbb{R}$$

2. *positive definiteness*

$$V[x(t)] > 0$$

3. *dissipativity*

$$\frac{d}{dt} V[x(t)] = \frac{\partial V}{\partial x} \frac{d[x(t)]}{dt} < 0$$

Note that the system dynamics is buried in the inequalities above, because both the positive definiteness and dissipativity is required along every possible trajectory in the state-space.

Theorem 7.3.1 (Lyapunov). *A system S is asymptotically stable (in the strong sense) if there exists a Lyapunov function with the above properties.*

Note that the Lyapunov criterion above is not constructive: it is the task of the user to find an appropriate Lyapunov function to show stability. Moreover, the reverse of the statement is *not* true.

Converse Lyapunov Theorem. While there is no general method of constructing Lyapunov functions for nonlinear systems, the existence of Lyapunov functions for asymptotically stable systems is theoretically guaranteed by the following well-known theorem.

Theorem 7.3.2 (Converse Lyapunov theorem). *Consider the following autonomous system*

$$\dot{x} = f(x) \tag{7.60}$$

where $x \in \mathbb{R}^n$, $f(0) = 0$, f is a smooth function on $\mathbb{R}^n \setminus \{0\}$ and continuous at $x = 0$. If the equilibrium $x = 0$ of the system is globally asymptotically stable, then there exists a positive definite and proper smooth function $V : \mathbb{R}^n \mapsto \mathbb{R}^+$ such that

$$\frac{\partial V}{\partial x} f(x) < 0, \quad \forall x \neq 0 \tag{7.61}$$

Note that if the stability of the equilibrium is not global, then the above theorem can still be applied to an appropriate neighborhood of $x = 0$.

7.3.2 Lyapunov Criterion for LTI Systems

The Lyapunov criterion is in a special "if-and-only-if" form for LTI system as follows:

Theorem 7.3.3 (Lyapunov for LTI systems). *The state matrix A of an LTI system is a stability matrix, i.e. $\text{Re}\{\lambda_i(A)\} < 0$ if and only if for any given positive definite symmetric matrix Q there exists a positive definite symmetric matrix P such that*

$$A^T P + PA = -Q \tag{7.62}$$

Note that Q can be positive semi-definite if the system is observable.

Proof. In order to show that the condition (7.62) is necessary, we define the *generalized energy* function of the system with realization (A, B, C) as

$$V[x(t)] = x^T Px > 0 \tag{7.63}$$

with P being positive definite.

It is now enough to show that the second Lyapunov property holds by computing the derivative

$$\frac{d}{dt} V[x(t)] = \frac{d}{dt}(x^T Px) = \dot{x}^T Px + x^T P\dot{x}$$

If we substitute the system equation $\dot{x} = Ax$ for $(A, 0, 0)$ in the equation above with $\dot{x}^T = (Ax)^T = x^T A^T$, we get

$$x^T A^T Px + x^T PAx = x^T (A^T P + PA)x = x^T (-Q)x$$

$A^T P + PA$ is negative definite if and only if Q is positive definite. For observable systems we can choose $Q = C^T C$.

The proof of the reverse direction is done by showing that if A is a stability matrix then

$$P = \int_0^\infty e^{A^T t} Q e^{At} dt \tag{7.64}$$

where Q is the given positive definite symmetric matrix, will satisfy $A^T P + PA = -Q$.

The following example illustrates the above theorem for a simple LTI system:

Example 7.3.1 (Lyapunov theorem for an LTI system)
Lyapunov theorem for a simple LTI system

Consider the following matrix:

$$A = \begin{bmatrix} -3 & 1 \\ 0 & -1 \end{bmatrix}$$

If possible, compute P from Equation (7.62) with

$$Q = \begin{bmatrix} 12 & 2 \\ 2 & 4 \end{bmatrix}$$

Since A is a stability matrix ($\lambda_1 = -3, \lambda_2 = -1, \text{Re}(\lambda_i) < 0$) and Q is positive definite (has positive eigenvalues), there is a positive definite symmetric matrix P to fulfill Equation (7.62). Searching for P in the form

$$P = \begin{bmatrix} p_{11} & p_{12} \\ p_{21} & p_{22} \end{bmatrix}$$

using the fact that $p_{12} = p_{21}$, we need to solve a linear set of equations to obtain the solution for P:

$$P = \begin{bmatrix} 2 & 1 \\ 1 & 3 \end{bmatrix}$$

Quadratically Stabilizable Systems. Note that we can try to apply the quadratic Lyapunov function (7.63) to any nonlinear system. For some nonlinear systems one can find a quadratic Lyapunov function $V(x) = x^T Q x$ with a suitably chosen Q. This gives rise to the definition below.

Definition 7.3.2 (Quadratically stabilizable systems)
A system is called quadratically stabilizable if there exists a static state feedback such that the closed-loop system is asymptotically stable with a quadratic Lyapunov function.

7.3.3 Lyapunov Criteria for LPV System Models

LPV system models have been introduced in Section 3.4 as mild extensions of LTI systems with varying parameters. The truncated version of an LPV state equation (see Equation (3.12)) used for asymptotic stability analysis is as follows:

$$\dot{x}(t) = A(\theta(t))x(t) \tag{7.65}$$

where the state matrix $A(\theta)$ is a function of a real-valued (possibly time-varying) parameter $\theta = [\theta_1, \ldots, \theta_k]^T$ from a bounded domain $\theta \in \Theta$.

If we look at LPV systems as being linear systems then we can expect them to be quadratically stabilizable with a suitable quadratic Lyapunov function

$$V(x) = x^T K x$$

where $K = K^T > 0$ is a symmetric positive definite matrix.

Definition 7.3.3 (Quadratic stability of LPV systems)
An LPV system with the truncated state equation (7.65) is said to be quadratically stable for perturbations Θ if there exists a matrix $K = K^T > 0$ such that

$$A(\theta)^T K + K A(\theta) < 0$$

for all perturbations $\theta \in \Theta$.

If we further assume that the state matrix $A(\theta)$ of the LPV system varies in a convex matrix polytope with corner-point matrices

$$\{A_1, A_2, \ldots, A_k\}$$

such that

$$A(\theta(t)) \in = Co\{A_1, A_2, \ldots, A_k\} := \{\sum_{i=1}^{k} \alpha_i A_i \; : \; \alpha_i \geq 0, \; \sum_{i=1}^{k} \alpha_i = 1\}$$

then a necessary and sufficient condition can be given for such a polytopic LPV system to be quadratically stable.

Theorem 7.3.4 (Quadratic stability of polytopic LPV systems). *If the LPV system (7.65) is a polytopic parameter-dependent model where $A(\theta) \in Co\{A_1, \ldots, A_k\}$ for all $\theta \in \Theta$, then it is quadratically stable if and only if there exists $K^T = K > I$ such that*

$$A_i^T K + K A_i < 0$$

for all $i = 1, \ldots, k$.

7.4 Stability of Process Systems

The asymptotic stability of process systems can be analyzed in three principally different ways depending on the process system in question:

- *local asymptotic stability analysis* can be performed after local linearization by investigating the eigenvalues of the resulting LTI state matrix A,
- *structural stability analysis* can be performed using the method of conservation matrices (later in Section 7.4.2),
- *global nonlinear stability analysis* can also be performed applying the Lyapunov theorem based on passivity analysis (see later in Section 8.5).

7.4.1 Structural Stability

The notion of structural dynamic properties relies on the concept of structure matrices of a dynamic state-space model.

Definition 7.4.1 (Structure matrix)
A structure matrix $[Y]$ of a real matrix Y is a matrix with structural zeros (denoted by 0) and structural non-zeros (denoted by $\times$) in its elements such that

$$[Y]_{ij} = \begin{cases} 0 & if \quad Y_{ij} = 0 \\ \times & otherwise \end{cases}$$

We may need to have the sign of the elements in the structure matrices, too. Signed structure matrices defined below serve this purpose.

Definition 7.4.2 (Signed structure matrix)
A signed structure matrix $\{Y\}$ of a real matrix Y is a matrix with structural zeros (denoted by 0) and signs (denoted by $-$ and $+$) in its elements such that

$$\{Y\}_{ij} = \begin{cases} 0 & if \quad Y_{ij} = 0 \\ + & if \quad Y_{ij} > 0 \\ - & if \quad Y_{ij} < 0 \end{cases}$$

The concept of structure matrices gives rise to recognition of classes of process models with the same structure and therefore similar dynamic properties.

Definition 7.4.3 (Structurally equivalent state-space models)
Two linear or linearized LTI systems with state-space representation matrices (A, B, C) and (A', B', C') are structurally equivalent if they give rise to the same set of signed structure matrices, i.e.

$$\{A\} = \{A'\}, \quad \{B\} = \{B'\}, \quad \{C\} = \{C'\}$$

With the notion of structurally equivalent state-space models we can say that a *structural property holds for a class of structurally equivalent state-space models if it holds for almost every member of the class (with the exception of a zero-measure sub-class).* This way we can define structural stability as follows:

Definition 7.4.4 (Structural stability)
A class of structurally equivalent state-space models is structurally asymptotically stable if almost every member in the class (with the exception of a zero-measure sub-class) is asymptotically stable.

The following example illustrates the notion of structural stability of a class in relationship to the stability of the members in the class.

Example 7.4.1 (Structural stability)
A simple structurally stable class

Let us consider the following state structure matrix $\{A\}$ defining a structurally equivalent LTI state-space model class:

$$\{A\} = \begin{bmatrix} - & 0 \\ 0 & - \end{bmatrix} \qquad (7.66)$$

The following two matrices are different numerical realizations of the above structure matrix:

$$A_1 = \begin{bmatrix} -2 & 0 \\ 0 & -3 \end{bmatrix}, \qquad A_2 = \begin{bmatrix} -0.05 & 0 \\ 0 & -1000 \end{bmatrix}$$

It is easy to see that both A_1 and A_2 are stability matrices, therefore the LTI systems they belong to are asymptotically stable.
On the other hand, every diagonal matrix with negative elements in its main diagonal is a stability matrix, because such matrices contain their real negative eigenvalues in their main diagonal. Therefore the whole LTI state-space model class defined by $\{A\}$ is structurally asymptotically stable.

7.4.2 Conservation Matrices

Conservation matrices are special stability matrices often obtained as state matrices for dynamic process models. Their stability property is used for proving structural stability of a class of process systems where the class members have a locally linearized state matrix with conservation matrix property irrespectively of the system parameters.

Definition 7.4.5 (Conservation matrix)
A real square matrix $F = \{f_{ij}\}_{i,j=1}^n$ of order n is said to be a column conservation matrix (or a row conservation matrix) if it is a matrix with dominant main diagonal with respect columns (or rows), i.e.

$$|f_{ii}| \geq \sum_{j \neq i} |f_{ij}| = R_i, \quad i = 1, 2, ..., n \qquad (7.67)$$

or

$$|f_{ii}| \geq \sum_{j \neq i} |f_{ji}| = C_i, \quad i = 1, 2, ..., n \qquad (7.68)$$

and its elements have the following sign pattern:

$$f_{ii} \leq 0, \quad f_{ij} \geq 0 , \ i \neq j \qquad (7.69)$$

In the case of proper inequality for every inequality in either Equation (7.67) or (7.68), F is said to be a a strict column conservation matrix or a strict row conservation matrix.

Conservation matrices are nonsingular and stable matrices [77], *i.e.* the real part of their eigenvalues is strictly negative for strict conservation matrices or non-positive otherwise.

7.5 Process System Examples

This section is devoted to showing how asymptotic stability analysis and structural stability analysis can be applied to process systems.

7.5.1 Stability Analysis of the Free Mass Convection Network

In this subsection we analyze the asymptotic stability of the free mass convection network described in Section 4.2.4 using the method of conservation matrices. The analysis is performed in two steps as follows:

1. First, we show that the convection matrix C_{conv} given by Equation (4.18) is a column conservation matrix.
2. Then we prove that the matrix product $C_{conv}\mathcal{K}$ in the state equation (4.22) is also a column conservation matrix with

$$\mathcal{K} = \text{diag}[\ \kappa^{(j)} \mid j = 1, \ldots, \mathcal{C}\ ]$$

being a diagonal matrix with positive elements.

With the program above, the analysis results are as follows:

1. We check the properties of conservation matrices for this case starting with the sign-pattern property and then investigating the dominant main diagonal property.
   - The sign pattern of the convection matrix C_{conv} given by Equation (4.18) is easy to check. As $\alpha_i^{(j)}$ is a ratio for which $0 \le \alpha_i^{(j)} \le 1$ holds, the convection matrix possesses the required sign property, that is,

   $$[C_{conv}]_{ii} = -(1 - \alpha_i^{(i)}) < 0$$

   $$[C_{conv}]_{ij} = \alpha_i^{(j)} \ge 0$$

   - The column-wise dominant main diagonal property is shown by using the defining Equation (4.16) for the flow ratios written in the form for the j-th column:

$$\sum_{i=1,\ i\neq j}^{c} \alpha_i^{(j)} + \alpha_j^{(j)} + \alpha_0^{(j)} = 1$$

Rearranging the equation and leaving out the non-negative term $\alpha_0^{(j)}$, we get

$$\sum_{i=1,\ i\neq j}^{c} \alpha_i^{(j)} \leq (1 - \alpha_j^{(j)})$$

which is exactly the inequality we wish to have.

2. Let us compute first the elements of the matrix $C_{conv}\mathcal{K}$ from their definitions to obtain

$$C_{conv}\mathcal{K} = \begin{bmatrix} -(1-\alpha_1^{(1)})\kappa^{(1)} & \alpha_1^{(2)}\kappa^{(2)} & ... & \alpha_1^{(C)}\kappa^{(C)} \\ ... & ... & ... & ... \\ \alpha_C^{(1)}\kappa^{(1)} & \alpha_C^{(2)}\kappa^{(2)} & ... & -(1-\alpha_C^{(C)})\kappa^{(C)} \end{bmatrix} \tag{7.70}$$

It can be seen that the matrix product $C_{conv}\mathcal{K}$ is obtained from C_{conv} in such a way that each of the columns of index j is multiplied by a positive constant $\kappa^{(j)}$ different for each column. Therefore both the sign property and the dominant main diagonal property remains unchanged, that is, the matrix $C_{conv}\mathcal{K}$ is also a column conservation matrix and as such is a stability matrix.

7.5.2 Lyapunov Function of the Free Mass Convection Network

As another technique for asymptotic stability analysis, we now construct a Lyapunov function for the LTI state-space model of the free convection network given in Equation (4.22).

First, we observe that the free mass convection network is a linear time-invariant system based on the state equation (4.22) with the following state and input variable vectors:

$$x_{FC} = M, \quad u_{FC} = V_{IN}$$

and state matrix

$$A_{FC} = C_{conv}\mathcal{K}$$

Therefore any function of the form

$$V_{FC}[x_{FC}(t)] = x_{FC}^T P x_{FC}$$

with P being a positive definite symmetric matrix, which is a solution of the equation

$$A_{FC}^T P + P A_{FC} = -Q$$

with Q being a positive definite symmetric matrix, is a Lyapunov function for the free mass convection network as is stated by Theorem 7.3.3.

Let us choose the simplest case when $Q = I$ then

$$P = -(A_{FC}^T + A_{FC})^{-1}$$

7.5.3 Structural Stability Analysis of Heat Exchanger Networks

We construct the dynamic process model of a heat exchanger network from sub-models of heat exchanger cells. Heat exchanger cells have been already introduced with their lumped process model developed in Section 4.4 before.

Engineering Model of a Heat Exchanger Cell in the Network. Recall, that a heat exchanger cell is a pair of regions connected by a heat-conducting wall. Assuming constant mass and constant physico-chemical properties in each balance volume, the engineering model of the j-th balance volume connected to the ℓ-th one forming jointly a heat exchanger cell is the potential form of the energy conservation balance equation for the temperature $T^{(j)}$ of the balance volume:

$$\frac{dT^{(j)}}{dt} = -\gamma^{(j)} v^{(j)} T^{(j)} + \gamma^{(j)} v^{(j)} T_B^{(j)} + K_{transfer}^{(j,\ell)} (T^{(\ell)} - T^{(j)}) \qquad (7.71)$$

where $K_{transfer}^{(j,\ell)} > 0$ is a constant transfer coefficient. Notice that only convective transport and a transfer term are present in the equation above, which can be seen as a generalized form of the intensive form conservation balances in Equations (4.47)–(4.48) with

$$\gamma^{(j)} = \frac{1}{V^{(j)}}, \quad K_{transfer}^{(j,\ell)} = \frac{U^{(j,\ell)} A^{(j,\ell)}}{V^{(j)}}$$

State Matrix of the Heat Exchanger Network. The set of state variables of the overall heat exchanger network is the set of temperatures in the cells, $i.e.$

$$\mathcal{X} = \bigcup_{j \in \mathcal{C}} T^{(j)} \qquad (7.72)$$

where the number of the balance volumes is $\mathcal{C} = 2N$, and N is the total number of heat exchanger cells in the heat exchanger network.

Let us rearrange the indices of the balance volumes forming the overall heat exchanger network so that the pair forming a heat exchanger cell are assigned indices next to each other, $i.e.$ $\ell = j+1$. Moreover, let us assume general connections between the cells described by Equations (4.17) and (4.16) in Section 4.2.4. Then the convective transport A^{Econv} and transfer $A^{Etransfer}$ state matrices forming the linearized state matrix A by

$$A = A^{Econv} + A^{Etransfer} \qquad (7.73)$$

are in the following form:

$$A^{Etransfer} = \begin{bmatrix} -K_{transfer}^{(1,2)} & K_{transfer}^{(1,2)} & \cdots & 0 & 0 \\ K_{transfer}^{(1,2)} & -K_{transfer}^{(1,2)} & 0 & \cdots & 0 \\ \cdots & \cdots & \cdots & \cdots & \cdots \\ 0 & \cdots & 0 & -K_{transfer}^{(2N-1,2N)} & K_{transfer}^{(2N-1,2N)} \\ 0 & \cdots & 0 & K_{transfer}^{(2N-1,2N)} & -K_{transfer}^{(2N-1,2N)} \end{bmatrix}$$

$$A^{Econv} = \begin{bmatrix} A_D^{(1)} & A_{OD}^{(1,2)} & \dots & A_{OD}^{(1,N)} \\ A_{OD}^{(2,1)} & A_D^{(2)} & \dots & A_{OD}^{(2,N)} \\ \dots & \dots & \dots & \dots \\ A_{OD}^{(N,1)} & A_{OD}^{(N,2)} & \dots & A_D^{(N)} \end{bmatrix} \qquad (7.74)$$

where the diagonal blocks in the convective matrix term are

$$A_D^{(j)} = \begin{bmatrix} -\gamma^{(j)}v^{(j)}(1-\alpha_j^{(j)}) & \gamma^{(j)}v^{(j)}\alpha_{j+1}^{(j)} \\ \gamma^{(j+1)}v^{(j+1)}\alpha_j^{(j+1)} & -\gamma^{(j+1)}v^{(j+1)}(1-\alpha_j^{(j+1)}) \end{bmatrix} \qquad (7.75)$$

Note that the diagonal blocks above are usually diagonal matrices because the convective connection between the cold and hot side of the same heat exchanger cell is a rare exception. Furthermore, the off-diagonal blocks $A_{OD}^{(i,j)}$ have elements being equal either to 0 or to a connection term $\gamma^{(j)}v^{(j)}\alpha_\ell^{(j)}$ where $\ell = i$ or $\ell = i+1$.

It is easy to see from the equations and matrices above that the convective transport state matrix A^{Econv} is a row conservation matrix while the transfer state matrix $A^{Etransfer}$ is a diagonal matrix with negative entries, therefore the overall state matrix is also a row conservation matrix acting as a stability matrix. *It implies that heat exchanger networks with constant mass and constant physico-chemical properties in each balance volume are structurally asymptotically stable* [34].

7.5.4 Structural Stability Analysis of a Binary Distillation Column with Constant Molar Flow and Vapor–liquid Equilibrium

The elementary engineering unit of a distillation column is a real or theoretical distillation tray which is a two-phase perfectly mixed subsystem with vapor and liquid phases. In the case of binary distillation, the composition of the binary mixture can be characterized by one mole fraction (*e.g.* the mole fraction of the more volatile component) in both the liquid (ξ) and the vapor (η) phases. The simple evaporator shown in Figure 4.1 in Section 4.3.2 can be regarded as a simple distillation tray.

Constant molar flows F for the liquid and V for the vapor phase are assumed with H_L being the liquid phase and H_V is the vapor phase molar hold-up – both of them are constants – on the tray.

Engineering Model of a Distillation Tray. If vapor–liquid equilibrium can be assumed on each tray and the vapor phase hold-up is negligible compared to the liquid phase hold-up, then the liquid and vapor phases of a tray form a balance volume in the dynamic model of the column with the liquid phase as the dynamic one. Moreover, because of the constant molar flow conditions, only the component conservation balance equation for the mole fraction of the more volatile component need be considered together with the equilibrium relationship. Therefore the original DAE model equations (4.25)–(4.45) specialize to the following set of differential equations:

$$H_L \frac{d\xi^{(j)}}{dt} = L(\xi^{(j+1)} - \xi^{(j)}) + V(\eta^{(j-1)} - \eta^{(j)}) \tag{7.76}$$

$$\eta^{(j)} = \mathcal{V}(\xi^{(j)}) \tag{7.77}$$

where

$$\Gamma^{(j)} = \frac{\partial \mathcal{V}(\xi^{(j)})}{\partial \xi^{(j)}} > 0 \tag{7.78}$$

for the j-th tray, *i.e.* for the j-th balance volume.

State Matrix of the Distillation Column. The set of state variables of the binary distillation column with constant molar flows is the set of liquid phase mole fractions on the trays, *i.e.*

$$\mathcal{X} = \bigcup_{j \in \mathcal{C}} \xi^{(j)} \tag{7.79}$$

where the number of the balance volumes is $\mathcal{C} = N$, and N is the number of the trays.

Substituting Equation (7.77) into the conservation balance equation (7.76) and linearizing the equations around the steady-state equilibrium points on the trays, the following state matrices result:

$$A^{Econv} = \begin{bmatrix} -\frac{L}{H_L} & \frac{L}{H_L} & 0 & \cdots & 0 \\ 0 & -\frac{L}{H_L} & \frac{L}{H_L} & \cdots & 0 \\ 0 & \cdots & \cdots & \cdots & 0 \\ \cdots & \cdots & 0 & -\frac{L}{H_L} & \frac{L}{H_L} \\ 0 & \cdots & \cdots & 0 & -\frac{L}{H_L} \end{bmatrix} \tag{7.80}$$

$$A^{Etransfer} = \begin{bmatrix} -\frac{V}{H_L}\Gamma^{(1)} & 0 & 0 & \cdots & 0 \\ \frac{V}{H_L}\Gamma^{(2)} & -\frac{V}{H_L}\Gamma^{(2)} & 0 & \cdots & 0 \\ 0 & \cdots & \cdots & \cdots & 0 \\ \cdots & \cdots & \frac{V}{H_L}\Gamma^{(N-1)} & -\frac{V}{H_L}\Gamma^{(N-1)} & 0 \\ 0 & \cdots & \cdots & \frac{V}{H_L}\Gamma^{(N)} & -\frac{V}{H_L}\Gamma^{(N)} \end{bmatrix} \tag{7.81}$$

It can be seen from the matrices above that they are both row conservation matrices, therefore *the binary distillation column with vapor–liquid equilibrium is structurally locally asymptotically stable* [34].

7.5.5 Structural Stability of a Binary Distillation Column with Constant Molar Flows in the Non-equilibrium Case

If equilibrium on the trays cannot be assumed, then another dynamic equation describing the component transport in the vapor phase should be considered instead of the equilibrium relationship (7.77). Therefore two balance volumes are needed to describe the behavior of a tray, one for the liquid and one for the vapor phase. The transfer between the phases is described by transfer rate expressions.

Engineering Model of the Distillation Tray. Because of the molar flow conditions, only the component transport equations for the mole fraction of the more volatile component in both phases have to be considered together with the equilibrium relationship

$$H_L \frac{d\xi^{(j)}}{dt} = L(\xi^{(j+1)} - \xi^{(j)}) + k_{Ltransfer}^{(j)}(\xi_*^{(j)} - \xi^{(j)}) \tag{7.82}$$

$$H_V \frac{d\eta^{(\ell)}}{dt} = V(\eta^{(\ell-1)} - \eta^{(\ell)}) + k_{Vtransfer}^{(\ell)}(\eta_*^{(\ell)} - \eta^{(\ell)}) \tag{7.83}$$

$$\eta_*^{(\ell)} = \mathcal{V}(\xi_*^{(j)}) \tag{7.84}$$

with the usual non-negativity condition (7.78) for the j-th liquid phase and for the ℓ-th vapor phase balance volumes on the tray and with

$$\Psi_{transfer}^{(j)} = k_{Ltransfer}^{(j)}(\xi_*^{(j)} - \xi^{(j)}) = -k_{Vtransfer}^{(\ell)}(\eta_*^{(\ell)} - \eta^{(\ell)}) = -\Psi_{transfer}^{(\ell)} \tag{7.85}$$

State Matrix of the Distillation Column. The set of state variables of the distillation column is the set of mole fractions in the liquid and vapor phase of the trays, $i.e.$

$$\mathcal{X} = \left(\bigcup_{j \in N} \xi^{(j)}\right) \bigcup \left(\bigcup_{j \in N} \eta^{(j)}\right) \tag{7.86}$$

where the number of balance volumes is $\mathcal{C} = 2N$, and N is the total number of trays in the column.

Let us rearrange the indices of the balance volumes forming the distillation column so that the pair forming a tray is assigned neighboring indices, $i.e.$ $\ell = j+1$. Moreover, let us assume that the mixture forms no binary azeotrope, therefore

$$k_{transfer}^{(j,\ell)} = k_{Ltransfer}^{(j)}, \quad k_{transfer}^{(\ell,j)} = k_{Vtransfer}^{(\ell)}, \quad \Gamma^{(j,\ell)} = \Gamma^{(j)} \tag{7.87}$$

Substituting Equation (7.84) into the conservation balances in Equations (7.82) and (7.83) and linearizing the equations around the steady-state equilibrium points on the trays, the following state matrices result:

$$
A^{Econv} = \begin{bmatrix}
-\frac{L}{H_L} & 0 & \frac{L}{H_L} & 0 & \cdots & 0 & 0 \\
0 & -\frac{V}{H_V} & 0 & 0 & \cdots & 0 & 0 \\
0 & 0 & \cdots & \cdots & \cdots & 0 & 0 \\
0 & \cdots & \cdots & 0 & 0 & -\frac{L}{H_L} & 0 \\
0 & \cdots & \cdots & 0 & \frac{V}{H_V} & 0 & -\frac{V}{H_V}
\end{bmatrix}
\tag{7.88}
$$

$$
A^{Etransfer} = \begin{bmatrix}
\Phi_1^{(1,2)} & \Phi_2^{(1,2)} & 0 & \cdots & & 0 \\
\Phi_1^{(2,1)} & \Phi_2^{(2,1)} & 0 & \cdots & & 0 \\
0 & \cdots & \cdots & \cdots & & 0 \\
\cdots & \cdots & 0 & \Phi_{2N-1}^{(2N-1,2N)} & \Phi_{2N}^{(2N-1,2N)} \\
0 & \cdots & \cdots & \Phi_{2N-1}^{(2N,2N-1)} & \Phi_{2N}^{(2N,2N-1)}
\end{bmatrix}
\tag{7.89}
$$

It can be seen from the matrices above that they both are row conservation matrices, therefore *the binary distillation column with no azeotropes is structurally locally asymptotically stable* [34].

7.6 Stability Analysis of a Simple Continuous Fermenter

The example of a simple continuous fermenter described in Section 4.5.3 is used here to show how stability analysis methods can be used to determine local stability near an optimal steady-state operating point and how the stability region can be determined [76].

The nonlinear state-space model of the fermenter is given in Equations (4.78)–(4.79). The variables and parameters of the fermentation process model are collected in Table 4.1.

7.6.1 Local Stability Analysis

The linearized state-space model of the continuous fermenter is also developed in Section 4.5.3 and is given in Equations (4.84)–(4.87).

A local stability analysis shows that the system is stable, within a neighborhood of the desired operating point, but because the point is very close to the fold bifurcation point

$$(X^*, S^*, F^*) = (4.8775, 0.2449, 3.2128)$$

this stability region is very small. This is illustrated in Figure 7.2, which shows that the system moves to the undesired wash-out steady state when it is started from close neighborhood of the desired operating point

$$X(0) = 4.7907\frac{g}{l}, \quad S(0) = 0.2187\frac{g}{l}$$

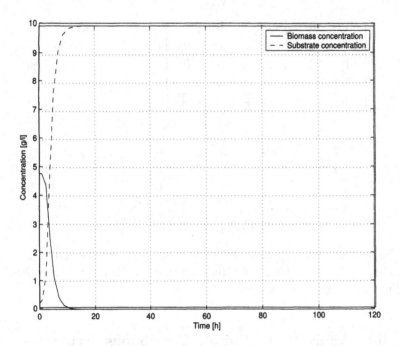

Figure 7.2. Open-loop behavior of the system

7.6.2 Stability Analysis Based on Local Linearization

The stability of the fermenter around the operating point depends upon the eigenvalues of the linearized state matrix A in Equations (4.84)–(4.87). These eigenvalues form a complex conjugated pair, in our case:

$$\lambda_{12} = -0.6017 \pm 0.5306i \tag{7.90}$$

We can see that the process is indeed locally stable around the operating point but the linear analysis does not provide any information on the extent of the stability region.

7.6.3 Nonlinear Stability Analysis

The nonlinear stability analysis is based on a Lyapunov technique which aims to find a positive definite scalar valued generalized energy function $V(x)$ which has a negative definite time derivative within the whole operating region. Most often a general quadratic Lyapunov function candidate is used in the form of

$$V(x) = x^T Q x$$

with Q being a positive definite symmetric quadratic matrix, which is usually diagonal. This function is scalar-valued and positive definite everywhere. The stability region of an autonomous nonlinear system is then determined by the negative definiteness of its time derivative:

$$\frac{dV}{dt} = \frac{\partial V}{\partial x}\dot{x} = \frac{\partial V}{\partial x}\bar{f}(x)$$

where $\bar{f}(x) = f(x)$ in the open loop-case (assuming zero input) and $\bar{f}(x) = f(x) + g(x)C(x)$ in the closed-loop case where $C(x)$ is the static linear or nonlinear feedback law.

The diagonal weighting matrix Q in the quadratic Lyapunov function is selected in a heuristic way: a state variable which does not produce overshoots during the simulation experiments gets a larger weight than another state variable with overshooting behavior. In the new norm defined by this weighting, a more accurate estimate of the stability region can be obtained.

With this analysis we cannot calculate the exact stability region but the results provide valuable information for selecting the controller type and tuning its parameters. The nonlinear stability analysis results in the time derivative of the quadratic Lyapunov function as a function of the state variables, which is a two-variable function in our case as seen in Figure 7.3. The stability region of the open-loop system is the region on the (x_1, x_2) plane over which the function is negative.

7.6.4 Stability Analysis based on an LPV Model

In order to show how the extension of the Lyapunov technique for linear parameter-varying (LPV) systems can be applied to a nonlinear system, we consider again the simple continuous fermenter model developed in Section 4.5.3 but we change the control input of the system to be the substrate feed concentration S_F.

The state-space model of the system is then given by the following equations:

$$\dot{X} = -\frac{F}{V}X + \mu(S)X \tag{7.91}$$

$$\dot{S} = -\frac{F}{V}S - \frac{1}{Y}\mu(S)X + \frac{F}{V}S_F \tag{7.92}$$

This can be written in LPV form as:

$$\frac{d}{dt}\begin{bmatrix} X \\ S \end{bmatrix} = \begin{bmatrix} \delta - \frac{F}{V} & 0 \\ -\frac{1}{Y}\delta & -\frac{F}{V} \end{bmatrix}\begin{bmatrix} X \\ S \end{bmatrix} + \begin{bmatrix} 0 \\ \frac{F}{V} \end{bmatrix}S_F \tag{7.93}$$

where $\delta = \mu(S)$ *is now considered as a time-varying parameter of the system.*

Now we construct a quadratic Lyapunov function candidate in the form of $V(x) = x^T P x$ and choose the positive definite matrix P such that it fulfills the conditions in Theorem 7.3.4.

Assume that δ is bounded, *i.e.* it takes its values in a compact interval:

$$\delta \in [0, \mu_{max}] \tag{7.94}$$

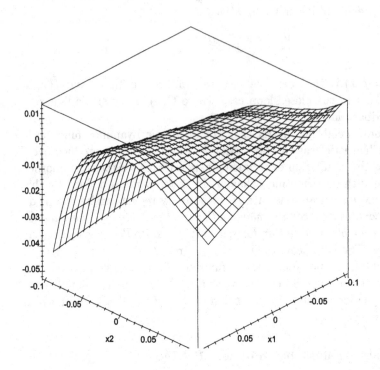

Figure 7.3. Time derivative of the Lyapunov function of the open-loop system with $Q = I$

The Lyapunov function of the simple continuous fermenter system model above has to fulfill the following linear matrix inequality (LMI) for any δ:

$$\begin{bmatrix} \delta - \frac{F}{V} & -\frac{1}{Y}\delta \\ 0 & -\frac{F}{V} \end{bmatrix} P + P \begin{bmatrix} \delta - \frac{F}{V} & 0 \\ -\frac{1}{Y}\delta & -\frac{F}{V} \end{bmatrix} < 0 \tag{7.95}$$

where "$<$" means "negative definite".

Since the LPV model is *affine* in δ, it is enough to investigate its stability at the end points of the interval [58]:

$$\begin{bmatrix} -\frac{F}{V} & 0 \\ 0 & -\frac{F}{V} \end{bmatrix} P + P \begin{bmatrix} -\frac{F}{V} & 0 \\ 0 & -\frac{F}{V} \end{bmatrix} = -2\frac{F}{V}P < 0 \tag{7.96}$$

$$\begin{bmatrix} \mu_{max} - \frac{F}{V} & -\frac{1}{Y}\mu_{max} \\ 0 & -\frac{F}{V} \end{bmatrix} P + P \begin{bmatrix} \mu_{max} - \frac{F}{V} & 0 \\ -\frac{1}{Y}\mu_{max} & -\frac{F}{V} \end{bmatrix} < 0 \tag{7.97}$$

Assume that

$$\mu_{max} < \frac{F}{V} \tag{7.98}$$

Then P will be searched in diagonal form:

$$P = \begin{bmatrix} p_1 & 0 \\ 0 & p_2 \end{bmatrix} \tag{7.99}$$

where $p_1, p_2 > 0$.

Since P is positive definite, the first inequality in Equation (7.96) is always fulfilled. The second one in Equation (7.97) is in the following form:

$$\begin{bmatrix} 2(\mu_{max} - \frac{F}{V})p_1 & -\frac{1}{Y}\mu_{max}p_2 \\ -\frac{1}{Y}\mu_{max}p_2 & -2\frac{F}{V}p_2 \end{bmatrix} < 0 \tag{7.100}$$

This inequality is fulfilled if and only if its elements fulfill the following conditions:

$$\det(M_{11}) = 2(\mu_{max} - \frac{F}{V})p_1 < 0 \tag{7.101}$$

$$\det(M) = -4(\mu_{max} - \frac{F}{V})\frac{F}{V}p_1p_2 - (\frac{1}{Y}\mu_{max}p_2)^2 > 0 \tag{7.102}$$

The first condition is always fulfilled if (7.98) holds, since $p_1 > 0$. The second condition is fulfilled if p_1 and p_2 are chosen as:

$$0 < p_2 < \frac{4Y^2F}{\mu_{max}^2 V} \left(\frac{F}{V} - \mu_{max} \right) p_1 = c\, p_1 \tag{7.103}$$

where c is positive according to the inequality in Equation (7.98).

The above derivation shows that the function

$$V(X, S) = p_1 X^2 + p_2 S^2 \tag{7.104}$$

is a Lyapunov function of the LPV system (7.93), if the conditions in Equations (7.98) and (7.103) are fulfilled.

7.7 Further Reading

The stability of linear systems is discussed in detail, e.g. in the excellent book [8]. [59] is a classical introductory book into the theory of Lyapunov's direct method. For the more mathematically oriented readers, [54] contains a wide arsenal of stability investigation methods and examples as well as a large number of further references.

7.8 Summary

Based on the stability notions and analysis tools developed for LTI systems, the basic concepts and algorithms for analyzing local and global stability of input-affine nonlinear systems are described in this section.

Besides the techniques that are based on linearized models and are suitable for investigating local stability near a steady-state operating point, the

Lyapunov method as the main global stability analysis tool is introduced here.

Structural stability analysis based on conservation matrices is also described in details.

The stability analysis methods are illustrated using case studies of process systems of practical importance: heat exchangers, distillation columns and continuous fermenters. Their stability analysis is based upon their input-affine nonlinear state-space model developed in Chapter 4.

7.9 Questions and Application Exercises

Exercise 7.9.1. Compare the local and global asymptotic stability analysis methods and the results of the analysis. Comment on the similarities and differences.

Exercise 7.9.2. What are the advantages and disadvantages of the method of conservation matrices in analyzing asymptotic stability of process systems?

Exercise 7.9.3. Consider the following two-dimensional nonlinear system model in its input-affine state-space form (the same as in Exercise 6.9.4):

$$\frac{dx_1}{dt} = x_1^{1/2} x_2^{3/2} - x_1 x_2^3 + x_1^{1/2} u_1$$

$$\frac{dx_2}{dt} = 3x_1 x_2^{1/2} - 2x_1^2 x_2$$

$$y = x_1^3 x_2^{2/3} - x_1^{1/2} x_2^2$$

Linearize the system model around a steady-state operating point.

Exercise 7.9.4. Consider the two-dimensional nonlinear system model above in Exercise 7.9.3.

Perform local stability analysis of the system using the state matrix A of the linearized system model around a steady-state operating point. (Use the result of Exercise 7.9.3.)

Exercise 7.9.5. Consider the following simple nonlinear truncated system model

$$\dot{x} = x(\lambda + ax)$$

on the open domain $x > 0$ with parameters $a > 0$, $\lambda > 0$.

Show that the function

$$V(x) = c\left(x - x^* - x^* \ln\left(\frac{x}{x^*}\right)\right)$$

is a Lyapunov function of the system with any $c > 0$ and $c^* > 0$ being the steady-state reference point of the system.

Exercise 7.9.6. Perform linear stability analysis of the nonlinear heat exchanger cell described in Section 4.4.4 using its linearized state-space model developed in Exercise 4.8.10.

Exercise 7.9.7. Perform structural stability analysis of the passive mass convection network described in Section 4.2.4.

Exercise 7.9.8. Analyze global stability of the passive mass convection network described in Section 4.2.4 using the Lyapunov function technique.

8. Passivity and the Hamiltonian View

Passivity and the Hamiltonian view on nonlinear systems were originally developed for mechanical systems and have proved to be useful in analyzing stability and designing stabilizing controllers. It is, however, possible and useful to extend these notions to other systems including process systems, and this is the subject of this chapter.

Passivity theory and the Hamiltonian view on nonlinear input-affine systems in general and on nonlinear process systems in particular are described here in the following sections:

- *The storage function and its properties*
  This section contains the mathematical definition and properties of the storage function together with its connection to the stability notions.
- *Passivity conditions and asymptotic stability*
  The definition of passive systems and the relation of passivity with Lyapunov's theorem on asymptotic stability is given here.
- *The Hamiltonian view*
  The Hamiltonian function is introduced as a general energy function using a mechanical analogy. The Euler-Lagrange equations are also developed and a relation between state and co-state variables are given.
- *Affine and simple Hamiltonian systems*
  An important special class of Hamiltonian systems, the so-called simple Hamiltonian systems, is introduced here for later use.
- *Passivity theory for process systems*
  The passivity theory is extended here for process systems. The storage function of process systems is given with its thermodynamical interpretation. Using the notion of passivity, a class of passive process systems is identified.
- *The Hamiltonian view on process systems*
  The Hamiltonian function is introduced for process systems and the thermodynamical meaning of the state and co-state variables is given based on the mechanical analogy.
- *Comparing passivity-based and Hamiltonian description*
  The two approaches available for describing nonlinear process systems are compared with respect to the modeling assumptions they require and the possibilities they offer for dynamic analysis.

- *Process system examples*
  Simple process system examples including a heat exchanger cell, a passive mass convection network and a simple unstable CSTR are given to illustrate the use of the tools and techniques introduced in this chapter.

8.1 The Storage Function and its Properties

The notion of L_2-gain of nonlinear systems introduced in Subsection 7.1.3 gives rise to the notion of available storage and then to the storage function. We recall that the notion of L_2-gain is related to BIBO stability and it is an extension of the finite gain concept for LTI systems. For the purpose of defining the storage function of a system, Equation (7.6) will be rearranged as follows:

Definition 8.1.1 (Available storage)
For all $x \in \mathcal{M}$, define the available storage $V_a : \mathcal{M} \to \mathbb{R}^+$ as:

$$V_a(x(T)) = \sup_u \frac{1}{2} \int_0^T \left(-\|y\|^2 + \gamma^2 \|u\|^2 \right) dt, \qquad u \in \mathcal{L}_2[0, T] \, , \; T > 0 \; (8.1)$$

given $x(0) = x_0$ and $\gamma \in \mathbb{R}$.

Observe that the defining equation above originates from Equation (7.6), which is regarded as the extension of the finite gain property to nonlinear systems with γ^2 being the static gain. In this context, $V_a(.)$ is seen as a supremum of the non-unique functions $K(.)$ given $\gamma^2 > 0$.

Definition 8.1.2 (Storage function)
The function $V : \mathcal{M} \to \mathbb{R}^+$ is called a storage function if the integral dissipation inequality is satisfied for any $t > t_0$:

$$V(x(t)) - V(x(t_0)) \leq \frac{1}{2} \int_{t_0}^t \left(\gamma^2 \|u\|^2 - \|y\|^2 \right) d\tau, \qquad V(0) = 0 \qquad (8.2)$$

for all $t > t_0$ and $u \in \mathcal{L}_2[t_0, t]$ with initial condition $x(t_0)$ and fixed t_0.

Note that *the storage function is not unique* for a given system. Moreover, $V(.)$ depends on the constant γ and on the choice of the input function u.

The integral on the right-hand side of inequality (8.2) represents the net "energy" feed into the system, that is, the difference between the feed at the input and the loss at the output. If we regard $V(x(t))$ as the energy content of the system at any time t, then its change over the interval $[t_1, t_2]$ is less than or equal to the net "energy" fed to the system if there is no internal energy source.

8.2 Passivity Conditions and Asymptotic Stability

This section shows how the above introduced storage function relates to asymptotic stability of a nonlinear system.

Definition 8.2.1 (Dissipative (passive) system)
The system is called dissipative (passive), with respect to the supply rate $\frac{1}{2}(\gamma^2\|u\|^2 - \|y\|^2)$, if the above storage function in Definition 8.1.2 exists.

Again, the definition above clearly depends on the constant γ and on the choice of the input function u.

Using the above concepts, we can easily arrive at the famous *Hamilton-Jacobi inequality*. First, we differentiate Equation (8.2) with respect to t and use the fact that we have a nonlinear system in input-affine form with $\dot{x} = f(x) + g(x)u$:

$$V_x(x)f(x) + V_x(x)g(x)u \leq \frac{1}{2}\gamma^2\|u\|^2 - \frac{1}{2}\|y\|^2, \qquad V(0) = 0 \tag{8.3}$$

for all $u \in \mathbb{R}^m$. Here $V_x = \text{grad } V$ is regarded as a *row* vector. Observe that the right-hand side of the above inequality is in the form:

$$\frac{1}{2}\gamma^2\|u\|^2 - \frac{1}{2}\|y\|^2 = \frac{1}{2}u^T u - \frac{1}{2}y^T y$$

We then substitute the output equation $y = h(x)$ into Equation (8.3) and perform the "completion of the squares" to get

$$V_x(x)f(x) - \frac{1}{2}\gamma^2(u - \frac{1}{\gamma^2}g^T(x)V_x^T(x))^T(u - \frac{1}{\gamma^2}g^T(x)V_x^T(x))$$
$$+ \frac{1}{2}\frac{1}{\gamma^2}V_x(x)g(x)g^T(x)V_x^T(x) + \frac{1}{2}h^T(x)h(x) \leq 0 \tag{8.4}$$

The above equation shows that the maximizing input is equal to

$$u^* = \frac{1}{\gamma^2}g^T(x)V_x^T \tag{8.5}$$

which permits us to obtain the supremum of the inequality (8.3). It is important to note that *the maximizing input u^* is in the form of a static nonlinear full state feedback* (see later in Chapter 9).

Finally, we substitute the maximizing input to Equation (8.4) and obtain the final form of the Hamilton-Jacobi inequality:

$$V_x(x)f(x) + \frac{1}{2}\frac{1}{\gamma^2}V_x(x)g(x)g^T(x)V_x^T(x) + \frac{1}{2}h^T(x)h(x) \leq 0 \tag{8.6}$$

$$V(0) = 0, \ x \in \mathcal{M}$$

Observe that the above equation relates the time derivative of the storage function with the gain condition under closed-loop conditions where we generate the maximizing input by a full, generally nonlinear, state feedback in Equation (8.5).

It follows from the definition of dissipative systems that the Hamilton-Jacobi inequality holds for dissipative systems controlled by the maximizing input. This means that we can check the dissipativity condition by checking the Hamilton-Jacobi inequality by using a candidate storage function $V(x)$. Also, we may try to use the inequality to construct a storage function for a given nonlinear system by solving it for V_x.

8.2.1 Lyapunov Functions and Storage Functions

There is a close connection between asymptotic (internal) stability of an input-affine nonlinear system described by the state-space model in Equation (7.5) in the form

$$\dot{x} = f(x) + g(x)u \qquad u \in \mathbb{R}^m, \ f(0) = 0$$
$$y = h(x) \qquad\qquad y \in \mathbb{R}^p, \ h(0) = 0$$

and the finiteness of L_2-gain. To see this we recall that for passive systems

$$\frac{d}{dt}V(x) \leq \frac{1}{2}(\gamma^2\|u\|_2^2 - \|y\|_2^2)$$

This means that the nonlinear system has a finite gain locally at each x, *i.e.* every time. As the supply rate is related to the local finite gain, this implies that the time derivative of the storage function is negative every time.

The *storage function* can then serve as a *Lyapunov function* [35]. The definition of the storage function implies that the storage function is positive definite and its time derivative is negative definite assuming that the supply rate is always non-negative (see Equation (8.2)).

8.3 The Hamiltonian View

The notion of Lagrangian and Hamiltonian systems (see, *e.g.* [80]) has been abstracted from the principles of theoretical mechanics when it has become clear that the underlying physics determines a strong special nonlinear structure which can be utilized effectively in nonlinear systems and control theory.

This section introduces the basic notions of the Hamiltonian description of nonlinear systems. The Hamiltonian view on both open-loop and closed-loop systems gives a relation between optimality, state and co-state variables. The relationship between the Hamiltonian description and the other tools and techniques applicable for input-affine nonlinear systems is also discussed.

8.3.1 Storage Function and the Hamiltonian View

It is possible to relate the above storage function concept in Section 8.1 to the Hamiltonian view. Equation (8.3) can be written as

$$K_\gamma\left(x, V_x^T(x), u\right) \leq 0, \qquad V(0) = 0 \qquad \text{for all } u \tag{8.7}$$

where K_γ is defined as

$$K_\gamma(x, p, u) = p^T f(x) + p^T g(x)u - \frac{1}{2}\gamma^2\|u\|^2 + \frac{1}{2}h^T(x)h(x) \tag{8.8}$$

where $p^T = (p_1, \ldots, p_n) = V_x$ denotes the so-called *co-state variable*. The following equivalent form

$$K_\gamma(x, p, u) = p^T \dot{x} - \frac{1}{2}(\gamma^2 u^T u - y^T y) = \frac{dV(x)}{dt} - \frac{1}{2}(\gamma^2 u^T u - y^T y) \tag{8.9}$$

shows that the function K_γ at any time t gives the difference between the differential stored energy (the first term) and the supply rate.

The function K_γ will serve as the Hamiltonian for both open-loop and closed-loop systems. Observe that K_γ depends on the gain γ^2 and the choice of the input, too. It is also important (see Equation (8.7)) that the Hamiltonian is negative definite for dissipative (passive) systems.

First, we examine the case of the supremum of the above function. Substituting the maximizing input

$$u^* = \frac{1}{\gamma^2}g^T(x)p \tag{8.10}$$

into Equation (8.8), we get the supremum of the Hamiltonian with a given gain γ^2 denoted by $H_\gamma(x, p) = K_\gamma(x, p, u^*)$:

$$H_\gamma(x, p) = p^T f(x) + \frac{1}{2}\frac{1}{\gamma^2}p^T g(x)g^T(x)p + \frac{1}{2}h^T(x)h(x) \tag{8.11}$$

8.3.2 The Hamiltonian Formulation in Classical Mechanics

In classical theoretical mechanics, energy is considered to be in two forms: potential energy V, which only depends on the position of the object ($V(q)$), and kinetic energy T, which depends on the velocity as well, *i.e.* $T(q, \dot{q})$. Then the following equations of motion can be used:

$$\frac{d}{dt}\left(\frac{\partial T}{\partial \dot{q}_i}\right) - \frac{\partial T}{\partial q_i} = F_i, \qquad i = 1, \ldots, n \tag{8.12}$$

where $\dim q = n$, q is the vector of generalized coordinates, $\dot{q}$ is the generalized velocity, $T(q, \dot{q})$ is the total *kinetic* energy, and F_i are the forces. The forces acting on the system can be external forces and internal forces derived from a potential $V(q)$:

$$F_i = -\frac{\partial V}{\partial q_i}(q) + F_i^e$$

Lagrangian Formulation. Define the Lagrangian function as

$$L_0(q, \dot{q}) = T(q, \dot{q}) - V(q) \tag{8.13}$$

where $V(q)$ is the *potential* energy. The Lagrangian function is used as a kind of "effect-function" where any motion is considered to take place according to the "minimal effect", *i.e.* in such a way that the trajectory $\{q(t), t_0 \leq t \leq t_1\}$ with t_0, t_1, $q(t_0)$ given is determined by

$$\arg\min_{q(t)} \int_{t_0}^{t_1} L_0(q, \dot{q}) dt$$

Substituting everything into the equations of motion, we arrive at the celebrated *Euler-Lagrange equations*:

$$\frac{d}{dt}\left(\frac{\partial L_0}{\partial \dot{q}_i}\right) - \frac{\partial L_0}{\partial q_i} = F_i^e, \qquad i = 1, \ldots, n \tag{8.14}$$

Control system equations are obtained from the above equations if the external forces are interpreted as control variables u_i, $i = 1, \ldots, n$:

$$\frac{d}{dt}\left(\frac{\partial L_0}{\partial \dot{q}_i}\right) - \frac{\partial L_0}{\partial q_i} = u_i, \qquad i = 1, \ldots, n \tag{8.15}$$

The usual equations of motion can be derived from Equation (8.15) considering that

$$T(q, \dot{q}) = \frac{1}{2}\dot{q}^T M(q)\dot{q}$$

with $M(q)$ being a positive definite matrix:

$$M(q)\frac{d^2q}{dt^2} + C(q, \dot{q}) + k(q) = Bu \tag{8.16}$$

where

$$k_i(q) = \frac{\partial V}{\partial q_i}(q), \quad B = \begin{bmatrix} I_m \\ 0 \end{bmatrix}$$

The above equation can be represented in a standard state-space model form as:

$$\frac{d}{dt}\begin{bmatrix} q \\ \dot{q} \end{bmatrix} = \begin{bmatrix} \dot{q} \\ M^{-1}(q)(C(q, \dot{q}) + k(q)) \end{bmatrix} + \begin{bmatrix} 0 \\ M^{-1}(q)B \end{bmatrix} u \tag{8.17}$$

Observe that the state vector of the system contains both the position and the velocity of the objects, *i.e.*

$$x = \begin{bmatrix} q \\ \dot{q} \end{bmatrix}$$

Definition 8.3.1 (Lagrangian for mechanical systems)

For a mechanical system with n degrees of freedom, in general, one can define a Lagrangian $L(q, \dot{q}, u)$ directly depending on the input u, and the equation of motion (8.15) becomes

$$\frac{d}{dt}\left(\frac{\partial L(q, \dot{q}, u)}{\partial \dot{q}_i}\right) - \frac{\partial L(q, \dot{q}, u)}{\partial q_i} = 0, \qquad i = 1, \dots, n \tag{8.18}$$

A direct comparison to (8.15) results in the relation:

$$L(q, \dot{q}, u) = L_0(q, \dot{q}) + \sum_{i=1}^{m} q_i u_i \tag{8.19}$$

where m is the number of inputs.

The resulted Lagrangian consists of two parts: the internal Lagrangian $L_0(q, \dot{q})$ and the coupling part describing the effect of the external forces.

A system described by Equation (8.18) is usually called a *Lagrangian control system* in the literature.

Hamiltonian Formulation. We can now move on to the Hamiltonian formulation. Define the generalized moments as

$$p_i = \frac{\partial L(q, \dot{q}, u)}{\partial \dot{q}_i}, \qquad i = 1, \dots, n \tag{8.20}$$

It can be shown that p_i, $i = 1, \dots, n$ are independent functions if the matrix $\{\partial^2 L/\partial \dot{q}_i \partial \dot{q}_j\}$ is nonsingular everywhere.

Definition 8.3.2 (Hamiltonian for mechanical systems)

The Hamiltonian function can be defined as the so-called Legendre transform of L:

$$H(p, q, u) = \sum_{i=1}^{n} p_i \dot{q}_i - L(q, \dot{q}, u) \tag{8.21}$$

It is simple to deduce that

$$\dot{q}_i = \frac{\partial H(p, q, u)}{\partial p_i}, \qquad i = 1, \dots, n \tag{8.22}$$

$$\dot{p}_i = -\frac{\partial H(p, q, u)}{\partial q_i}, \qquad i = 1, \dots, n \tag{8.23}$$

The first equation holds by construction and the second one follows by substituting Equation (8.20) into Equation (8.18). The second equation is then the equation of motion in a Hamiltonian formulation.

Note that the variables p_i, $i = 1, \dots, n$, being the moments of the system, are called *co-state variables*. Unlike the original state variables q_i, $i = 1, \dots, n$ being co-ordinates, the co-state variables are *extensive quantities which are conserved*.

The system described by (8.22)–(8.23) is called a *Hamiltonian control system* in the literature.

Hamiltonian Function as the Total Energy. There is another interpretation of the Hamiltonian function. If $L(q, \dot{q}, u)$ is defined by Equation (8.19), then it follows that

$$H(q, p, u) = H_0(q, p) - \sum_{i=1}^{m} q_i u_i \tag{8.24}$$

where H_0 is the Legendre transform of L_0. Observe that the Hamiltonian function contains two parts: the internal Hamiltonian $H_0(q, p)$ and the coupling part.

For mechanical systems, L_0 is given as $L_0 = T(q, \dot{q}) - V(q)$, where

$$T(q, \dot{q}) = \frac{1}{2} \dot{q}^T M(q) \dot{q} \quad \text{and} \quad p = \partial L / \partial \dot{q}$$

Therefore

$$p = M(q) \dot{q} \tag{8.25}$$

holds, with $M(q)$ being a positive definite matrix. Then it follows that

$$H_0(q, p) = \frac{1}{2} p^T M^{-1}(q) p + V(q) \tag{8.26}$$

that is, the *internal energy* of the system contains a kinetic and a potential energy part, respectively.

The following simple mechanical example illustrates the above:

Example 8.3.1 (Hamiltonian model of a spring)
Harmonic oscillation

Let us have an ideal spring with spring constant k connected to a mass point m. Describe the motion around its steady-state point at $x = 0$.

The Lagrange function is in the form:

$$L = T - V = \frac{1}{2} m \dot{x}^2 - \frac{1}{2} k x^2$$

Therefore the co-state variable is

$$p = \frac{\partial L}{\partial \dot{x}} = m \dot{x}$$

Then the Hamiltonian is in the form of

$$H = T + V = \frac{p^2}{2m} + \frac{kx^2}{2}$$

and the canonical equations of motion are

$$\left(\frac{\partial H}{\partial p} =\right) \dot{x} = \frac{p}{m}, \quad \left(\frac{\partial H}{\partial x} =\right) \dot{p} = -kx$$

8.4 Affine and Simple Hamiltonian Systems

The notion of Hamiltonian systems has been introduced already in Section 8.3. Two special cases of Hamiltonian systems are introduced here for later use.

8.4.1 Affine Hamiltonian Input–output Systems

We remember that a system described by Equations (8.22)–(8.23) is called a *Hamiltonian system* in the literature. In some of the cases, the Hamiltonian function is in a special, *input-affine* form

$$H(q, p, u) = H_0(q, p) - \sum_{j=1}^{m} H_j(q, p) u_j \tag{8.27}$$

where H_0 is the *internal Hamiltonian* and $H_j(q,p)$, $j = 1, \ldots, m$ are the *interaction or coupling Hamiltonians*, q is the state and p is the co-state variable.

Note that the above Hamiltonian is an extension to the one introduced in theoretical mechanics in such a way that the original interaction terms $\sum_{j=1}^{m} q_j u_j$ have been extended to $\sum_{j=1}^{m} H_j(p, q) u_j$.

Furthermore, we may associate *natural outputs* to the system as follows:

$$y_j = -\frac{\partial H}{\partial u_j}(q, p, u) = H_j(q, p), \quad j = 1, \ldots, m \tag{8.28}$$

Notice that in the original definition of the Hamiltonian control system there were no output variables present. The natural outputs are *artificially* defined new output variables to the system, even their number is made equal to the number of input variables m.

With the above definitions, *the nonlinear state-space model of an affine Hamiltonian input–output system* is in the form:

$$\dot{q}_i = \frac{\partial H_0}{\partial p_i}(q,p) - \sum_{j=1}^{m} \frac{\partial H_j}{\partial p_i}(q,p)\, u_j, \quad i = 1,\dots,n \tag{8.29}$$

$$\dot{p}_i = -\frac{\partial H_0}{\partial q_i}(q,p) + \sum_{j=1}^{m} \frac{\partial H_j}{\partial q_i}(q,p)\, u_j, \quad i = 1,\dots,n \tag{8.30}$$

$$y_j = H_j(q,p), \quad j = 1,\dots,m \tag{8.31}$$

Note again that these definitions are extensions of the original mechanically motivated ones.

8.4.2 Simple Hamiltonian Systems

We may further specialize the above definition to obtain the notion of simple Hamiltonian systems.

Definition 8.4.1 (Simple Hamiltonian system)
A simple Hamiltonian system is an affine Hamiltonian system where the functions H_0, H_1, ..., H_m are of the following special form:

$$H_0(q,p) = \frac{1}{2}p^T G(q)p + V(q) \tag{8.32}$$

with $G(q)$ being a positive definite symmetric $n \times n$ matrix for every q, and

$$H_j(q,p) = H_j(q), \quad j = 1,\dots,m \tag{8.33}$$

With the above requirements for the internal and coupling Hamiltonian, *the nonlinear state-space model of an affine Hamiltonian input–output system* specializes to the form:

$$\dot{q}_i = \frac{\partial H_0}{\partial p_i}(q,p) = G(q)p, \quad i = 1,\dots,n \tag{8.34}$$

$$\dot{p}_i = -\frac{\partial H_0}{\partial q_i}(q,p) + \sum_{j=1}^{m} \frac{\partial H_j}{\partial q_i}(q)\, u_j \tag{8.35}$$

$$y_j = H_j(q), \quad j = 1,\dots,m \tag{8.36}$$

The above simple Hamiltonian systems are special cases of the original, mechanically motivated ones in two main points:

- $H_0(p,q)$
  is in a special form suitable for both mechanical and process systems.
- $H_j(q)$
  is again special depending only on the co-state variable.

The following simple mechanical example, a modified version of the system introduced in Example 8.3.1, illustrates the above.

Example 8.4.1 (Hamiltonian model of a forced spring)
Simple Hamiltonian description of a harmonic oscillation with external force

Let us have an ideal spring with spring constant k connected to a mass point m and let us have an external force F^e acting on the mass. Give the simple Hamiltonian description of the system.

We describe the motion around its steady-state point at $q_0 = 0$ and regard the external force as the input

$$u = F^e$$

The Lagrange function is in the form:

$$L = T - V + uq = \frac{1}{2}m\dot{q}^2 - \frac{1}{2}kq^2 + uq$$

Therefore the co-state variable is

$$p = \frac{\partial L}{\partial \dot{q}} = m\dot{q}$$

Then the Hamiltonian is in the form of

$$H = H_0 - H_1 u = T + V - H_1 u = \frac{p^2}{2m} + \frac{kq^2}{2} - qu \qquad (8.37)$$

and the canonical equations of motion are

$$\left(\frac{\partial H}{\partial p} = \right) \dot{q} = \frac{p}{m}, \quad \left(-\frac{\partial H}{\partial q} = \right) \dot{p} = -kq + u \qquad (8.38)$$

It follows from the state equations (8.38) and from the Hamiltonian (8.37) that the natural output to the system is

$$y = H_1(q) = q \qquad (8.39)$$

that is, the spring system with an external force as its input is a simple Hamiltonian system.

If we consider friction in the example above, which acts as a damping factor, then we can still obtain a simple Hamiltonian model of the system as is shown in the next example.

Example 8.4.2 (Hamiltonian model of a damped spring)
Simple Hamiltonian description of damped harmonic oscillation with external force

Develop the Hamiltonian description of the ideal spring with a mass point and with friction as a dissipation external force $F^e = \mu mg$. Use the description of the ideal case in Example 8.3.1.

The state variable q of this system is the distance from the equilibrium point at $q_0 = 0$. The Lagrangian is in the form:

$$L = T - V = \frac{1}{2}m\dot{q}^2 - \frac{1}{2}kq^2 + \mu mgq$$

Therefore the co-state variable is

$$p = \frac{\partial L}{\partial \dot{q}} = m\dot{q}$$

Then the Hamiltonian is in the form:

$$H = T + V = \frac{p^2}{2m} + \frac{kq^2}{2} - \mu mgq \qquad (8.40)$$

and the canonical equations of motion are

$$\left(\frac{\partial H}{\partial p} =\right) \dot{q} = \frac{p}{m}, \quad \left(-\frac{\partial H}{\partial q} =\right) \dot{p} = -kq + \mu mg \qquad (8.41)$$

8.5 Passivity Theory for Process Systems: a Lagrangian Description

Passivity analysis of a nonlinear system requires us to have a storage function. This section is devoted to show how a suitable storage function can be developed based on thermodynamical principles.

The system variables and the general form of state-space models of process systems introduced in Section 4.2 is the starting point of the analysis. We recall that the state equations originate from conservation balances for the canonical set of conserved extensive quantities (mass, component masses and energy) in each balance volume.

8.5.1 System Variables

In order to get a compact description, we use a notation for the conserved extensive quantities in the balance volume j, assuming that they have been arranged into a vector:

$$\phi^{(j)} = [\; m_1^{(j)} \;\; \ldots \;\; m_K^{(j)} \;\; E^{(j)} \;]^T \tag{8.42}$$

Then the vector of the corresponding thermodynamical potentials for the j-th region is defined as

$$P^{T(j)} = \frac{\partial S^{(j)}}{\partial \phi^{(j)}} \tag{8.43}$$

and is in the form

$$P^{(j)} = [\; -\frac{\mu_1^{(j)}}{T^{(j)}} \;\; \ldots \;\; -\frac{\mu_K^{(j)}}{T^{(j)}} \;\; \frac{1}{T^{(j)}} \;]^T \tag{8.44}$$

where $\mu_k^{(j)}$ is the chemical potential for the component k and $T^{(j)}$ is the temperature in region j.

Gibbs-Duhem Relationship. The above entries of $P^{(j)}$ are the consequence of the Gibbs-Duhem relationship, which is valid under the standard assumption of local thermodynamical equilibrium [27] applied everywhere in each balance volume j. The Gibbs-Duhem relationship for open systems out of equilibrium is of the form

$$S^{(j)} = P^{T(j)}\phi^{(j)} + \left(\frac{p}{T}\right)^{(j)} V^{(j)} \tag{8.45}$$

where p is the pressure. From homogenity of the entropy as a function of the other extensive variables $(\phi^{(j)}, V^{(j)})$, Equation (8.45) can be written in differential form as follows:

$$dS^{(j)} = \left(\frac{1}{T}\right)^{(j)} dE^{(j)} + \left(\frac{p}{T}\right)^{(j)} dV^{(j)} - \sum_{k=1}^{K} \left(\frac{\mu_k}{T}\right)^{(j)} dm_k^{(j)} \tag{8.46}$$

where the operator d stands for total derivative. Note that the second term is equal to zero, because $dV^{(j)} = 0$ caused by the "general modeling assumptions" in Section 4.1.1.

The Reference Point. The notion of *centered variables*

$$\overline{\varphi} = \varphi - \varphi^*$$

with φ^* being the reference point of the variable φ introduced in Section 4.2.2, plays an important role in the following. *It is important to note that the reference point for the conserved extensive quantities should be a thermodynamical equilibrium point in the general case. This implies that the reference point for the input variables should then be chosen in such a way that the right-hand side of the general conservation balances (4.3) gives zero in their reference point.*

There are a number of consequences of the requirement above on the selection of the reference points:

1. If there is *no source term* in the general conservation balances then the requirement specializes to
$$u^* = 0$$

2. If we have *no transfer term* then we can select u^* arbitrarily. Then it is reasonable to choose q^* such that it is equal to its setpoint value for control and then choose u^* to satisfy the requirement.

3. In the general case we find a thermodynamical equilibrium point close to the reference points and then choose u^* from it.

Extensive-intensive Relationship. It is important to note that there is a simple linear relationship between conserved extensive quantities and their potentials in the form

$$\overline{P}^{(j)} = Q^{(j)}\overline{\phi^{(j)}} \tag{8.47}$$

where $Q^{(j)}$ is a negative definite square matrix for which the following relationship holds:

$$Q_{k\ell}^{(j)} = \frac{\partial^2 S^{(j)}}{\partial \phi_k^{(j)} \partial \phi_\ell^{(j)}} \tag{8.48}$$

The above relationship is a consequence of the concave nature of the system entropy $S^{(j)}$ of the j-th region and of the fact that the system entropy is a first-order homogeneous function of the conserved extensive quantities.

8.5.2 Thermodynamical Storage Function

The storage function proposed in [29] originates from the overall system entropy of the following general form:

$$V = \sum_{j=1}^{\mathcal{C}} \left(-\overline{S}^{(j)} + \sum_{k=1}^{K+1} P_k(\phi_k^{*(j)})\overline{\phi}_k^{(j)} \right) = \sum_{j=1}^{\mathcal{C}} \left(P^T(\phi^{*(j)})\overline{\phi}^{(j)} - \overline{S}^{(j)} \right) \quad (8.49)$$

where $S^{(j)}$ is the entropy and $P^{T(j)}$ is the vector of thermodynamical potentials for the conserved extensive quantities $\phi^{(j)}$ in balance volume j.

It can be shown that the above function is convex and positive semidefinite, taking its minimum at the reference equilibrium (steady-state) point.

We can transform V to a more suitable form, taking into account the infinite series expansion of the entropy $S^{(j)}$ around its equilibrium value $S^{*(j)}$:

$$S^{(j)} = S^{*(j)} + P^T(\phi^{*(j)})\overline{\phi}^{(j)} + \overline{\phi}^{T(j)}Q^{(j)}\overline{\phi}^{(j)} + \mathcal{O}(\overline{\phi}^{(j)3}) \quad (8.50)$$

with $Q^{(j)}$ being a negative definite symmetric matrix (the Hessian matrix of the entropy function)

$$Q_{k\ell}^{(j)} = \frac{\partial^2 S^{(j)}}{\partial \phi_k^{(j)} \partial \phi_\ell^{(j)}} \quad (8.51)$$

to get

$$V = \sum_{j=1}^{\mathcal{C}} \left(-\overline{\phi}^{T(j)}Q^{(j)}\overline{\phi}^{(j)} - \hat{V}^{(j)}(\overline{P}^{(j)}) \right) \quad (8.52)$$

where $\hat{V}(.)$ is a nonlinear function.

It is important to observe that we can get the extensive–intensive relationship in Equation (8.47) from the Taylor series expansion of the defining relationship of the thermodynamical potentials P in Equation (8.43) around the equilibrium (steady-state) point with the same matrix $Q^{(j)}$ as in Equation (8.51):

$$P^{(j)}(\phi^{(j)}) = P^{(j)}(\phi^{*(j)}) + \left.\frac{\partial P^{(j)}}{\partial \phi^{(j)}}\right|_* \left(\phi^{(j)} - \phi^{*(j)} \right) + \mathcal{O}(\overline{\phi}^{(j)2})$$

taking into account only the first-order (linear) terms. For this, we observe that

$$\left.\frac{\partial P^{(j)}}{\partial \phi^{(j)}}\right|_* = Q^{(j)}$$

is the Hessian matrix of the entropy function, the same as defined in Equation (8.51).

8.5.3 Transfer Terms

The difference between the thermodynamical potentials as driving force variables induces transfer flows between regions in mutual contact. It is assumed that the transfer term of any of the regions can be decomposed into pair-wise transfers, which are additive:

$$q_{\phi,transfer}^{(j)} = \sum_{\ell=1}^{c} \psi_{q,transfer}^{(j,\ell)}$$

Onsager Relationship. The celebrated *Onsager relationship* of the irreversible thermodynamics connects the transfer fluxes $\psi_{\phi,transfer}^{(j,\ell)}$ with the potentials in the form

$$q_{\phi,transfer}^{(j,\ell)} = L^{(j,\ell)} \left(P^{(j)} - P^{(\ell)} \right) \tag{8.53}$$

where the matrix $L^{(j,\ell)}$ is positive definite and symmetric [27], [42].

It is important to note that usually the difference of the engineering driving forces $(\Phi^{(j)} - \Phi^{(\ell)})$ is considered on the right-hand side of the transfer term (8.53). It can be shown, however, that the positive definiteness and symmetricity of the coefficient matrix cannot be guaranteed in this case, therefore we shall always use the thermodynamically correct expression in Equation (8.53) instead. The following table shows the relation between conserved extensive, engineering and thermodynamical driving force variables.

Table 8.1. Relation between thermodynamical variables

Case	Conserved extensive	Engineering driving force	Thermodynamical driving force
general ($\overline{P} = Q\overline{\phi}$)	ϕ	Φ	P
energy ($\overline{E} = c_P m \overline{T}$)	energy E	temperature T	rec. temperature $\frac{1}{T}$
comp. mass ($\overline{\mu_k} = RT \ln c_k$)	comp. mass m_k	concentration c_k	rel. chem. potential $-\frac{\mu_k}{T}$

If we further *choose the thermodynamical equilibrium point as a reference value for the thermodynamical potentials* where

$$P^{*(j)} = P^{*(\ell)} \tag{8.54}$$

then we can develop the normalized version of the Onsager relation in the form of

$$q_{\phi,transfer}^{(j,\ell)} = L^{(j,\ell)} \left(\overline{P}^{(j)} - \overline{P}^{(\ell)} \right), \quad L^{(j,\ell)} > 0, \quad \left(L^{(j,\ell)} \right)^T = L^{(j,\ell)} \tag{8.55}$$

8.5.4 Decomposition of the Time Derivative of the Storage Function

The time derivative of the storage function above in Equation (8.49) is in the form

$$\frac{d\mathcal{V}}{dt} = -\sum_{j=1}^{\mathcal{C}}\sum_{i=1}^{K+1} \overline{P}_i^{(j)} \frac{d\overline{\phi}_i^{(j)}}{dt} = -\sum_{j=1}^{\mathcal{C}} \overline{P}^{T(j)} \frac{d\overline{\phi}^{(j)}}{dt} \tag{8.56}$$

Observe that the equation above contains the left-hand side of the normalized general conservation balance equation (4.3).

With the above observation, the time derivative of the storage function can be decomposed in exactly the same way as for the right-hand side of the conservation balances

$$\frac{d\mathcal{V}}{dt} = \mathcal{V}_t^{transfer} + \mathcal{V}_t^{iconv} + \mathcal{V}_t^{oconv} + \mathcal{V}_t^{source} \tag{8.57}$$

where the terms have the following special form:

1. The convective and transfer terms of the balance equation (8.57) above is in the form of

$$\mathcal{V}_t^{oconv} + \mathcal{V}_t^{transfer} = x^T R_{XX} x \tag{8.58}$$

   with R_{XX} being a negative definite matrix in the case of Kirchoff or passive convection between the balance volumes.
2. The inflow term is in the form

$$\mathcal{V}_t^{iconv} = -\sum_{j=1}^{\mathcal{C}} v^{(j)} \alpha_0^{(j)} \overline{P}^{T(j)} \overline{\phi}_{in}^{(j)} \tag{8.59}$$

   where $\alpha_0^{(j)}$ is the ratio of the system inflow to the overall input flow rate of the j-th region.

8.5.5 Passivity Analysis

The passivity analysis is performed by using the decomposition of the time derivative of the storage function in Equation (8.57), *assuming that the conserved extensive quantities at the inlet are chosen as input variables:*

$$u_j = \overline{\phi}_{in}^{(j)}$$

If every term on its right-hand side is negative definite then the system itself is passive, that is, asymptotically stable.

The special form of the convective and transfer terms in Equation (8.58) shows that *the transfer and convective mechanisms (assuming Kirchhoff or passive convection) are passive in every process system.*

In the case of *open-loop systems*, we assume that the inlet inflow terms in any of the regions are kept constant, that is,

$$\overline{\phi}_{in}^{(j)} = 0$$

This implies that the inlet inflow term $\mathcal{V}_t^{iconv}$ is identically zero, that is, it gives no addition to the time derivative and is therefore "neutral" from the viewpoint of the passivity of the system.

Finally, the source term $\mathcal{V}_t^{source}$ is indefinite (neither positive nor negative definite) in the general case. This means that:

1. An open-loop process system with no source and only Kirchhoff or passive convection is always passive, that is, asymptotically stable [29].
2. A suitable nonlinear feedback can be designed to make the sum of the terms $\mathcal{V}_t^{iconv}$ and $\mathcal{V}_t^{source}$ always negative definite (see later in Chapter 12).

8.6 The Hamiltonian View on Process Systems

In order to find the Hamiltonian function of a process system, we use the mechanical analog described in Section 8.3 and extend it to process systems using the storage function (8.49) [30].

First, we notice that the Hamiltonian for mechanical systems is the overall energy of the system and therefore it can serve as a storage function for their passivity analysis. In process systems the direction of the "movement" of the system, that is, its time evolution in the state-space is driven by entropy: closed systems are in equilibrium when their overall entropy is maximum.

8.6.1 State, Co-state and Input Variables

Following the mechanical analog, the following state and co-state variables are identified for the Hamiltonian description of process systems:

$$p = [\overline{\phi}^{T(1)} \ \cdots \ \overline{\phi}^{T(C)}]^T \tag{8.60}$$

$$q = [\overline{P}^{T(1)} \ \cdots \ \overline{P}^{T(C)}]^T \tag{8.61}$$

This means that the generalized moments are the normalized conserved extensive variables and the generalized co-ordinates are the normalized thermodynamical driving forces.

Static Relationship between States and Co-states. In addition to the state-space model equations, there is a linear static (*i.e.* time-invariant) relationship originating from Equation (8.48) between the state and co-state variables of a process system in the form

$$q = Qp \tag{8.62}$$

where Q is a negative definite symmetric block-diagonal matrix of the form

$$Q = \begin{bmatrix} Q^{(1)} & 0 & \cdots & 0 \\ 0 & Q^{(2)} & 0 & \cdots & 0 \\ \cdots & \cdots & \cdots & \cdots \\ 0 & 0 & \cdots & Q^{(C)} \end{bmatrix} \tag{8.63}$$

The above equations are the consequence of the definitions of the state and co-state variables in Equations (8.60) and (8.61), and of the concavity of the entropy function near an equilibrium (steady-state) point.

Onsager Relationship. There is another relationship between the state and co-state variables of a process system which can be derived from the form of the Onsager relationship in Equation (8.55), which gives an expression for the transfer rate of conserved extensive quantities as a function of the related thermodynamical potentials as driving forces:

$$\dot{p}_{transfer} = \psi_{\phi,transfer} = \mathcal{L}\, q, \quad \mathcal{L} > 0, \; \mathcal{L}^T = \mathcal{L} \tag{8.64}$$

where the matrix $\mathcal{L}$ is positive definite and symmetric in the following form:

$$\mathcal{L} = \frac{1}{2} \sum_{j=1}^{c} \sum_{\ell=1}^{c} \left(I^{(j,\ell)} \otimes L^{(j,\ell)} \right)$$

Observe that the matrix $\mathcal{L}$ in (8.64) is the same as the transfer matrix $A_{transfer}$ in the decomposed state equation (4.15).

8.6.2 Input Variables for the Hamiltonian System Model

In order to have a Hamiltonian description of process systems, we have to assume that only the mass flow rates form the set of input variables, with the inlet engineering driving force variables being constant. Then the set of normalized input variables is the same as in Equation (4.12):

$$u = [(\bar{v}^{(j)}), \; j = 1, \ldots, C]^T$$

8.6.3 The Hamiltonian Function

Following the mechanical analog, the Hamiltonian function of process systems should describe the direction of changes taking place in an open-loop system. The mechanical analog and the general defining properties (8.32)–(8.36) of simple Hamiltonian systems will be used for the construction of the simple Hamiltonian model of process systems by using the Onsager relationship (8.64) and the conservation balances (4.3) together with the relationships between the state and co-state variables (8.62).

Construction of the Hamiltonian. The Hamiltonian function is then constructed in two sequential steps as follows:

1. *The kinetic term*
   The kinetic term is constructed from the Onsager relationship (8.64) by transforming it into the form in Equation (8.34) by using the relationship between the state and co-state variables in Equation (8.62) to obtain

$$\dot{q} = (Q\mathcal{L}Q)\,p = \mathcal{G}p \tag{8.65}$$

where $\mathcal{G}$ is a positive semi-definite symmetric matrix that does not depend on q. Symmetricity follows from the identity

$$(Q\mathcal{L}Q)^T = Q\mathcal{L}Q \ \text{ with } \ Q^T = Q, \ \ \mathcal{L}^T = \mathcal{L}$$

and positive semi-definiteness is a simple consequence of the positive semi-definiteness of $\mathcal{L}$.

$$x^T\,(Q\mathcal{L}Q)\,x = (Qx)^T\,\mathcal{L}\,(Qx) = y^T\mathcal{L}y \ \geq \ 0 \ \forall\, x$$

The kinetic term $T(p)$ in the Hamiltonian function will be constructed to satisfy Equation (8.34), *i.e.*

$$T(p) = \frac{1}{2}p^T\mathcal{G}p \tag{8.66}$$

2. *The potential term and the coupling Hamiltonians*
   The potential term and the coupling Hamiltonians are derived by matching the terms in the special form of the defining Hamiltonian property (8.35), taking into account that now $\mathcal{G}$ does not depend on q

$$\dot{p}_i = -\frac{\partial H_0}{\partial q_i}(q,p) + \sum_{j=1}^{m}\frac{\partial H_j}{\partial q_i}(q)\,u_j = -\frac{\partial V(q)}{\partial q_i} + \sum_{j=1}^{m}\frac{\partial H_j}{\partial q_i}(q)\,u_j \tag{8.67}$$

with the terms in the decomposed general conservation balances (4.15) with the flow rates as input variables:

$$\dot{p} = (\mathcal{L}q + Q_\phi(q)) + \sum_{j=1}^{m} \left(N_j Q^{-1} q u_j + B_{j,conv} u_j \right) \tag{8.68}$$

$$= f(q) + \sum_{j=1}^{m} g_j(q) u_j$$

where $B_{j,conv}$ is the j-th column of the input convection matrix B_{conv}. From this correspondence the potential energy term $V(q)$ and the coupling Hamiltonians $H_j(q)$ should satisfy

$$f_i(q) = -\frac{\partial V(q)}{\partial q_i}, \quad g_{ij}(q) = \frac{\partial H_j}{\partial q_i}(q) \tag{8.69}$$

With the above ingredients the Hamiltonian function of a process systems is written in the form

$$H(p, q, u) = T(p) + V(q) - \sum_{j=1}^{m} H_j(q) \, u_j \tag{8.70}$$

satisfying all the required properties in Equations (8.34)–(8.36).

We may further specialize the form of the Hamiltonian function above by using the decomposition of the state function $f(q)$ in Equation (8.68) according to the mechanisms (transfer and source) to get a decomposition of the potential term $V(q)$ to satisfy (8.69):

$$V(q) = V_{transfer}(q) + V_Q(q)$$

$$V_{transfer}(q) = -\frac{1}{2} q^T \mathcal{L} q, \quad \frac{\partial V_Q(q)}{\partial q} = -Q_\phi(q) \tag{8.71}$$

Substituting the above decomposed potential term to the Hamiltonian function (8.70) above, we obtain

$$H(p, q, u) = V_Q(q) - \sum_{j=1}^{m} H_j(q) \, u_j \tag{8.72}$$

It can be seen that *there is no kinetic term in the equation above because*

$$T(p) = -V_{transfer}(q)$$

but the internal Hamiltonian does only contain the potential term originating from the sources.

The above constructive derivation gives rise to the following theorem, which summarizes the main result:

Theorem 8.6.1. *Given a process system with the input variables (4.12) as the flow rates, the state and co-state variables (8.60) and (8.61) enables us to construct a simple Hamiltonian system model with the Hamiltonian function (8.72) and with the underlying relationships in Equations (8.65), (8.68) and (8.62), (8.69).*

Remarks. There are important remarks to the above as follows:

- *The supply rate*
  As has been mentioned before, the internal Hamiltonian function is a storage function of the system with respect to the supply rate $\sum_{j=1}^{m} H_j(q)\, u_j$. Recall that the so-called natural output variables have been defined to be equal to the coupling Hamiltonians, that is, $y_j = H_j(q)$. Therefore the supply rate of the system can be written in the form of

$$a = y^T u$$

- *The internal Hamiltonian*
  It is important to note that the internal Hamiltonian above is a storage function for process systems which gives the total entropy power (entropy produced in unit time). This can be checked by computing the units in the Hamiltonian function. Therefore *the storage function derived from the simple Hamiltonian description is entirely different from the entropy-based storage function in Equation (8.49)*. This is explained by the known fact that the storage function of a nonlinear system is *not unique*.

- *The time derivative of the Hamiltonian storage function*
  For stability analysis the time derivative of the storage function is important, which is in a special form in this case:

$$\frac{dH_0(q,p)}{dt} = \sum_{i=1}^{n} \left(\frac{\partial H_0}{\partial p_i} \frac{dp_i}{dt} + \frac{\partial H_0}{\partial q_i} \frac{dq_i}{dt} \right)$$

$$= \sum_{i=1}^{n} \left(\frac{\partial H_0}{\partial p_i} \frac{\partial H_0}{\partial q_i} - \frac{\partial H_0}{\partial q_i} \frac{\partial H_0}{\partial p_i} \right) = 0$$

where the defining Equations (8.34) and (8.35) have been used for the derivation with $u = 0$.

8.7 Comparing the Hamiltonian and Lagrangian Description for Process Systems

The process systems' engineering conditions that enable us to construct a Lagrangian or a Hamiltonian description are compared in Table 8.2 below. It is shown that *process systems with constant mass hold-ups in each balance volume, constant physico-chemical properties and no source enable both conditions if the input variables are suitably chosen.*

8.8 Simple Process Examples

Simple process examples serve to illustrate the notions and tools described in this chapter. The heat exchanger cell, the free mass convection network and a simple unstable CSTR model developed in Chapter 4 is used.

Table 8.2. Comparison of the Hamiltonian and Lagrangian system models

	Lagrangian	**Hamiltonian**
Input	inlet intensive quantities	mass flow rates
Internal mechanism	transfer, (source) convection	transfer, (source)
Special condition		**Hamiltonian source**

8.8.1 Storage Function of the Heat Exchanger Cell

The storage function of the nonlinear heat exchanger cell model developed in Section 4.4 is constructed from its state-space model by identifying its state and input variables.

System Variables. The vector of *conserved extensive quantities* of the nonlinear state-space model Equations (4.47)–(4.48) is composed of the vector of internal energies E_h and E_c for the two balance volumes in question:

$$\phi = [E_h, E_c]^T, \quad n = 2 \tag{8.73}$$

that is, the regions $j = h, c$ of the hot and cold sides respectively.

Note that with our modeling assumptions, the following relation holds between the volume-specific energy and the temperature of a balance volume:

$$E_j = c_{Pj}\rho_j V_j T_j + E_0 \tag{8.74}$$

with E_0 being an arbitrary reference.

Centered Variables and their Relationships. Let us consider a reference state E_h^*, E_c^* defined as a stationary (equilibrium) state with the corresponding driving forces $\frac{1}{T_h^*}, \frac{1}{T_c^*}$ and let us denote the difference from the reference state by overlining, *i.e.*

$$\overline{E}_j = E_j - E_j^* \ .$$

The special form of the Onsager relation between the volume-specific energy and the reciprocal temperatures as driving forces can be derived from the relation in Equation (8.74) by expanding $\frac{1}{T^{(j)}}$ into a Taylor series to obtain

$$\frac{\overline{1}}{T_j} = M_j'\overline{E}_j, \quad M_j' = -\frac{1}{c_{Pj}\rho_j V_j \left(T_j^*\right)^2} \tag{8.75}$$

with $M_j' < 0$ being a constant in this case.

Storage Function. The storage function $\mathcal{V}(\phi^{(j)})$ for the j-th region of the simple heat exchanger can be defined as follows:

$$\mathcal{V}(E_j) = \left(-\overline{S}_j + \frac{1}{T_j^*} \overline{E}_j \right) \qquad (8.76)$$

where $\overline{S}_j$ is the difference between the entropy of the balance volume characterized by the thermodynamical state E_j and the reference entropy for $j = h, c$.

With the relation between the potential and conserved extensive variables in Equation (8.75), the storage function takes the form:

$$\mathcal{V}(E_j) = \left(-\overline{S}_j + \frac{1}{T_j^*} \frac{1}{M_j'} \frac{\overline{1}}{T_j} \right) \qquad (8.77)$$

The storage function of the overall heat exchanger composed of the two $(j = h, c)$ regions is simply the sum of the individual storage functions (8.77) above:

$$\mathcal{V}(\phi) = \sum_{j=h,c} \mathcal{V}(E_j) \qquad (8.78)$$

depending on the non-normalized vector ϕ of conserved extensive quantities in Equation (8.73) of the overall system.

8.8.2 Hamiltonian Description of the Heat Exchanger Cell

The Hamiltonian system model of the heat exchanger cell is again developed from its state-space model described in Section 4.4 by identifying its state, co-state and input variables.

The continuous time state equations of the heat exchanger cell are derived from the following energy conservation balances:

$$\frac{dE_c(t)}{dt} = v_c(t) c_{Pc}(T_{ci}(t) - T_c(t)) + UA(T_h(t) - T_c(t)) \qquad (8.79)$$

$$\frac{dE_h(t)}{dt} = v_h(t) c_{Ph}(T_{hi}(t) - T_h(t)) + UA(T_c(t) - T_h(t)) \qquad (8.80)$$

where T_{ji} and T_j are the inlet and outlet temperature and v_j is the mass flow rate of the two sides $(j = c, h)$ respectively. Note that we have now two regions, that is, $C = 2$. Observe that the model equations above contain an input and output convection and a transfer term expressed in engineering driving forces, but there is no source term.

System Variables. The vector of conserved extensive quantities consists of the internal energies for the two regions:

$$\phi = [E_h, E_c]^T, \quad n = 2$$

Let us choose the volumetric flow rates v_c and v_h as input variables.

Extensive–intensive Relationship. Let us choose a reference thermodynamical equilibrium state for the heat exchanger cell such that

$$T_h^* = T_c^* = T^* \tag{8.81}$$

The energy–temperature relations are known from elementary thermodynamics:

$$E_j = c_{Pj} m_j T_j + E_0, \quad \overline{E}_j = E_j - E_j^* \tag{8.82}$$

where m_j is the constant overall mass of region $j = h, c$. From Equations (8.82), we obtain by expanding $\frac{1}{T^{(j)}}$ into the Taylor series that

$$\overline{T}_j = -(T^*)^2 \frac{\overline{1}}{T_j}$$

$$\frac{\overline{1}}{T_j} = Q^{(j)} \overline{E}_j, \quad Q^{(j)} = -\frac{1}{c_{Pj} m_j (T^*)^2}$$

with $Q^{(j)} < 0$ being a constant in this case. Therefore

$$Q = \begin{bmatrix} -\frac{1}{c_{Ph} m_h (T^*)^2} & 0 \\ 0 & -\frac{1}{c_{Pc} m_c (T^*)^2} \end{bmatrix}$$

Centered System Variables. With the reference equilibrium state (8.81), we can easily define the centered state and thermodynamical potential (driving force) variables as

$$q = [\ \frac{\overline{1}}{T_h}\ , \ \frac{\overline{1}}{T_c}\ ]^T$$

$$p = [\ \overline{E}_h\ , \ \overline{E}_c\ ]^T$$

From the conservation balance equations (i.e. Equations (8.79)–(8.80)), it follows that now the reference point for the input variables is the zero vector, therefore

$$u = [\ v_h,\ v_c\ ]^T$$

Decomposed State Equation in Input-affine Form. With the above defined centered system variables the conservation balance equations, which are Equations (8.79)–(8.80), can be written in the following canonical form:

$$\frac{dp}{dt} = \mathcal{A}_{transfer}q + B_{1c}u + \sum_{i=1}^{2} N_i u_i \qquad (8.83)$$

with

$$\mathcal{A}_{transfer} = (T^*)^2 \begin{bmatrix} UA & -UA \\ -UA & UA \end{bmatrix}, \quad B_{1c} = \begin{bmatrix} c_{Ph}\overline{T}_{hi} & 0 \\ 0 & c_{Pc}\overline{T}_{ci} \end{bmatrix} \qquad (8.84)$$

$$N_1(\overline{T}_h) = \begin{bmatrix} -c_{Ph}\overline{T}_h \\ 0 \end{bmatrix}, \quad N_2(\overline{T}_c) = \begin{bmatrix} 0 \\ -c_{Pc}\overline{T}_c \end{bmatrix}$$

where $\mathcal{A}_{transfer}$ and B_{1c} are constant matrices and N_1 and N_2 are linear functions of the centered engineering driving force variables. Observe that now the transfer function matrix $L^{(c,h)}$ is just a constant UA (a 1×1 matrix) and

$$I^{(c,h)} = \begin{bmatrix} 1 & -1 \\ -1 & 1 \end{bmatrix}$$

Hamiltonian System Model. Let us now develop the Hamiltonian description of the nonlinear heat exchanger cell model.

The internal Hamiltonian of the heat exchanger cell system is easily constructed from the special form of the Hamiltonian developed for process systems in Equation (8.72), taking into account that there is no source term in the conservation balance equations, *i.e.* $V_Q(q) = 0$

$$H_0(q,p) = 0 \qquad (8.85)$$

Now we need to identify the coupling Hamiltonians $H_1(q)$ and $H_2(q)$ from the co-state equation (8.83) for the input variables

$$u = [\, v_h, \ v_c \,]^T$$

The coupling Hamiltonians can be reconstructed from the vector functions $g_1(q)$ and $g_2(q)$ respectively, which are the gradient vectors of the corresponding coupling Hamiltonians:

$$\frac{\partial H_1}{\partial q} = g_1(q), \quad \frac{\partial H_2}{\partial q} = g_2(q)$$

Observe that the gradients are naturally given in terms of the engineering driving force variables Φ but we need to transform them into the form depending on the state variables q. In the heat exchanger cell case, we have

$$g_1(\overline{T}_h) = \begin{bmatrix} -c_{Ph}\overline{T}_h + c_{Ph}\overline{T}_{hi} \\ 0 \end{bmatrix} = \begin{bmatrix} c_{Ph}\left(T^*\right)^2 q_1 + c_{Ph}\overline{T}_{hi} \\ 0 \end{bmatrix} = g_1(q)$$

$$g_2(\overline{T}_h) = \begin{bmatrix} 0 \\ c_{Pc}\overline{T}_c + c_{Pc}\overline{T}_{ci} \end{bmatrix} = \begin{bmatrix} 0 \\ -c_{Pc}\left(T^*\right)^2 q_2 + c_{Pc}\overline{T}_{ci} \end{bmatrix} = g_2(q)$$

By partial integration, we obtain

$$H_1(q) = \frac{1}{2}c_{Ph}\left(T^*\right)^2 q_1^2 + c_{Ph}\overline{T}_{hi}q_1 \tag{8.86}$$

$$H_2(q) = \frac{1}{2}c_{Pc}\left(T^*\right)^2 q_2^2 + c_{Pc}\overline{T}_{ci}q_2 \tag{8.87}$$

From the passivity analysis we know that the system is inherently passive but it has a pole at the stability boundary because there is no source term and $V_Q(q) = 0$. Therefore we can perform stabilization by a derivative feedback and loop-shaping by a static feedback using PD controllers (see later in Subsection 12.1.3).

8.8.3 Hamiltonian Description of the Free Mass Convection Network

The engineering dynamic model of the free mass convection network (see in Section 4.2.4) originates from the conservation balance equations of the overall mass given in Equation (4.22).

The linear state-space model (4.22) is transformed into its Hamiltonian canonical form by considering the following set of state, co-state and input variables:

$$q = [\, \rho g\overline{h}^{(1)}, \ldots, \rho g\overline{h}^{(C)} \,]^T$$
$$p = [\, \overline{V}^{(1)}, \ldots, \overline{V}^{(C)} \,]^T$$
$$u = [\, v_{IN}^{(1)}, \ldots, v_{IN}^{(C)} \,]^T$$

where $\overline{h}^{(j)}$ is the centered level of the liquid in balance volume j where constant cross-section A and liquid density ρ is assumed in each balance volume and the gravitation constant is denoted by g. There are simple static relationships between the level $h^{(j)}$ and volume $V^{(j)}$ of the j-th balance volume

$$V^{(j)} = Ah^{(j)} \,.$$

Then the balance equations written for the co-state variables become

$$\frac{dp}{dt} = C_{conv}\mathcal{K}p - \frac{1}{\rho}V_{IN} = C_{conv}\mathcal{K}Q^{-1}q - \frac{1}{\rho}V_{IN} \tag{8.88}$$

Now the matrix Q is in the form of

$$Q = \begin{bmatrix} \frac{\varrho g}{A} & 0 & \cdots & 0 \\ 0 & \frac{\varrho g}{A} & \cdots & 0 \\ \cdots & \cdots & \cdots & \cdots \\ 0 & 0 & \cdots & \frac{\varrho g}{A} \end{bmatrix}$$

Finally, the internal Hamiltonian of the passive mass convection system is

$$H_0(q, p) = T(p) + V(q) = q^T \left(C_{conv} \mathcal{K} Q^{-1} \right) q \qquad (8.89)$$

which only has a kinetic term.

8.8.4 Hamiltonian Description of a Simple Unstable CSTR

Let us have an isotherm CSTR with fixed mass hold-up m and constant physico-chemical properties where a second-order

$$2A + S \rightarrow T + 3A$$

autocatalytic reaction takes place. The engineering model of the reactor is developed in Section 4.5.1 from the conservation balance for the component mass of component A.

Assume that the inlet concentration of component A (c_{Ain}) is constant and the inlet mass flow rate v is used as an input variable. We develop the Hamiltonian description of the system around its steady-state point determined as the setpoint for passivation and loop-shaping. This description is then used later in Section 12.4 for a nonlinear proportional feedback controller to stabilize the system.

Conservation Balance Equation and System Variables. The state equation is a single component mass conservation balance equation for component A in the form

$$\frac{dm_A}{dt} = \frac{d(m \cdot c_A)}{dt} = vc_{Ain} - vc_A + k \cdot m \cdot c_A^2 \qquad (8.90)$$

where m_A is the component mass, m is the constant overall mass and k is the reaction rate constant. Note that we only have a single balance volume, therefore $C = 1$.

The given steady-state concentration c_A^* with a nominal mass flow rate v^* satisfies

$$0 = v^*(c_{Ain} - c_A^*) + k \cdot m \cdot (c_A^*)^2$$

From this, we can determine v^* as

$$v^* = -\frac{k \cdot m \cdot (c_A^*)^2}{c_{Ain} - c_A^*}$$

which should be non-negative, therefore $c_{Ain} \leq c_A^*$ should hold. The given steady-state concentration c_A^* also determines the nominal value of the conserved extensive quantity m_A, the component mass in this case being:

$$m_A^* = m \cdot c_A^*$$

The engineering driving force variable to the component mass m_A is the concentration c_A and the thermodynamical potential (driving force) variable is

$$\overline{P} = -R' \overline{\mu}_A = -R'' \overline{\ln c_A} \cong -R\overline{c}_A$$

with R being a constant under isotherm conditions and assuming ideal mixtures.

Hamiltonian System Model. It follows from the above that the centered system variables for the Hamiltonian description of the simple unstable CSTR are as follows:

$$p = \overline{m}_A = m_A - m_A^*, \quad q = -(c_A - c_A^*), \quad u = v - v^* \tag{8.91}$$

Observe that the constant R has been omitted from the definition of the co-state variable q as compared to the thermodynamical driving force q above. From the variable definitions above we see that the matrix Q specializes to

$$Q = -\frac{1}{m}$$

From the conservation balance equation in Equation (8.90), it is seen that there is only a single balance volume present in the system and there is no transfer term. Therefore the transfer coefficient matrix $\mathcal{L} = 0$. This implies that now the reference point for the state and co-state variables can be chosen arbitrarily. However, the source term Q_ϕ is now present as a second-order term originating from the autocatalytic second-order reaction.

If we substitute the centered variables in Equation (8.91) to the conservation balance equation (8.90), the following normalized state equation is obtained:

$$\frac{dp}{dt} = \left(k \cdot m \cdot q^2 - (2k \cdot m \cdot c_A^* + v^*) \cdot q \right) + (\overline{c}_{Ain} + q)u \tag{8.92}$$

From the equation above we can identify the elements of the Hamiltonian description to be

$$\frac{\partial V_Q}{\partial q}(q) = -\left(k \cdot m \cdot q^2 - (2k \cdot m \cdot c_A^* + v^*) \cdot q \right), \quad \frac{\partial H_1}{\partial q}(q) = (\overline{c}_{Ain} + q)$$

By partial integration we obtain

$$V_Q(q) = -\frac{1}{3} k \cdot m \cdot q^3 + \frac{1}{2} (v^* + 2k \cdot m \cdot c_A^*) \cdot q^2 \tag{8.93}$$

$$H_1(q) = \frac{1}{2} q^2 + \overline{c}_{Ain} q \tag{8.94}$$

Passivity Analysis of the Unstable CSTR. The passivity analysis is performed by using the internal Hamiltonian of the system

$$H_0(q,p) = V_Q(q) = -\frac{1}{3} k \cdot m \cdot q^3 + \frac{1}{2} (v^* + 2k \cdot m \cdot c_A^*) \cdot q^2$$

We can see that the above function is of no definite sign because of the presence of the second- and third-order terms of different constant coefficients and different signs. This means that the system fails to be passive in the general case.

8.9 Further Reading

The idea of using physical knowledge to construct storage functions for passivity analysis is not new: it is worth reading the early papers and books, such as [80] and [19].

The traditional and still main application area of Lagrangian and Hamiltonian system models and their physics-based control is in mechatronics (robots): see [65] for a simple introductory paper. A good early survey of Hamiltonian system models in mechatronics is found in [52].

Process Systems. There is a wide and growing literature in the field of making connections between thermodynamics, variational calculus and the theory of Hamiltonian systems. Ydstie [83] offers a recent survey of related papers. The papers in this area can be classified according to the system models they use for dynamic analysis and controller design.

- *Lagrangian process system models*
  The idea of investigating dissipation and passivity of process systems based on thermodynamical principles was introduced in the 1990's [55], [84], [23], where its implications of controller design have also been explored. The approach is applied to networks of process systems, *i.e.* to composite process systems with several balance volumes in [28].
- *Hamiltonian process system models*
  The principles of constructing a Hamiltonian system model can be found in [30].

8.10 Summary

Lagrangian and Hamiltonian system models are special, extended model forms for input-affine nonlinear systems. Besides the usual elements of nonlinear input-affine state-space models, they have an additional structure in their set of system variables. Moreover, they are equipped with an additional storage function which enables us to carry out passivity analysis in a straight-forward way.

It is shown that process systems with constant mass hold-ups in each balance volume, constant physico-chemical properties and no source enable both Lagrangian and Hamiltonian system models if the input variables are suitably chosen. The selection of the input variables changes the effect of convection from an internal mechanism in the Lagrangian case to a coupling one in the Hamiltonian case. Besides the special input selection, an additional source condition should hold for Hamiltonian process systems.

8.11 Questions and Application Exercises

Exercise 8.11.1. What is the relationship between passivity and asymptotic stability of nonlinear systems? Comment also on the relationship between Lyapunov functions and storage functions.

Exercise 8.11.2. What are the ingredients (elements) of a simple Hamiltonian system model? Comment on the relationship between a usual input-affine state-space model and a Hamiltonian system model. What are the common and additional elements in a Hamiltonian model?

Exercise 8.11.3. Give the elements of a Lagrangian system model of a process system. Comment on the relationship between a usual input-affine state-space model and a Lagrangian process system model. What is the engineering meaning of the additional model element in a Lagrangian systems model?

Exercise 8.11.4. What are the elements of a Hamiltonian system model of a process system? Give the relationship between a usual input-affine state-space model and a Hamiltonian process system model. What is the engineering meaning of the additional model elements in a Hamiltonian systems model of a process system?

Exercise 8.11.5. Compare the elements of the Lagrangian and Hamiltonian system models in the example of the heat exchanger cell which is described in Section 8.8.1 with a Lagrangian and in Section 8.8.2 with a Hamiltonian system model.

Exercise 8.11.6. Construct a Lagrangian system model of the simple unstable CSTR model developed in Subsection 4.5.1 using the inlet concentration c_{Ain} of component A as the input variable. Compare the resulting model with the Hamiltonian one developed in Subsection 8.8.4.

Exercise 8.11.7. Consider a simple unstable CSTR model developed in Subsection 4.5.1 but with the reaction rate expression

$$r = k \cdot m \cdot c_A^4$$

Construct a Hamiltonian system model of the CSTR following the derivation in Subsection 8.8.4.

Exercise 8.11.8. Consider the input-affine nonlinear state-space model of a simple continuous fermenter that is developed in Section 4.5.3 and given in Equations (4.78)–(4.79). Construct a simple Hamiltonian model for this system.

9. State Feedback Controllers

This chapter summarizes the basic notions and techniques for the control of nonlinear systems. The material is mainly a revised summary of the control of LTI systems ([39], [9]) with a view to extending them for the nonlinear case.

The material in the chapter is broken down into the following sections:

- *The notion of control and feedback*
  We start with the notion of control and the tasks we need to perform for control. The general notion of feedback together with that of output and state feedback is also given here, and the notion of full state feedback and linear and nonlinear static feedback is introduced.
- *Pole-placement controller for LTI systems*
  One of the basic types of full state feedback controllers is the pole-placement controller, which is described here for LTI systems.
- *LQR for LTI systems*
  The linear quadratic regulator (LQR) is a widely applied and theoretically important controller. It is introduced here for LTI systems and its properties are shown.
- *Hamiltonian view on controlled systems*
  Here we connect the optimization-type formulation of controller design with a Hamiltonian system description in order to facilitate the understanding and analysis of the properties of controlled systems.
- *Case study: linear control of a continuous fermentation process*
  In the example of a simple continuous fermentation process, this section shows the common way of controlling nonlinear process systems using linear controllers and indicates possible problems associated with this approach.

9.1 Control and Feedback

Until now, we investigated process systems as they are: we have learned about various mathematical descriptions of systems (*representation of systems*) and analyzed dynamical properties like stability, controllability and observability (*system analysis*). We did this with a purpose: the main aim of system representation and analysis is to influence the behavior of systems (that is, to *control* them) and to make decisions based on their expected behavior (to perform a *diagnosis* on them).

9.1.1 Control and Optimization

When we manipulate the input signals of a system in order to achieve some goal concerning its behavior then we "control" it. From this description, it follows that we should have a *goal* or *control aim* in mind when approaching a system with the intention of controlling it.

A suitable or the best input signal can then be selected by a simple "generate and test" method: we try an input signal, evaluate it from the viewpoint of the goal and then try another one, possibly by improving the previous one. The way we figure out a suitable or the best input signal to achieve our goal is usually performed by *optimization*.

Aim. Control methods can be classified according to the *control aim*, or in other words, the goal function. The most common goals are as follows:

1. We want to *keep a prescribed value of the output signals as close as possible* against disturbances which cause them to change. This control task is called *regulation*. Linear Quadratic Regulators (LQRs) perform control according to this type of goal function.
2. We may want to *move the state variables of the system from one prescribed initial state to a given final state as quickly as possible*. This control task is called *time-optimal control* and it is quite common in process systems.
3. We might try to *stabilize* unstable systems or improve their dynamic behavior. This is done by *stabilizing control*, for example by pole-placement control.

Process Models as Constraints for Control. The control goals are formulated in terms of a desired output behavior (*e.g.* constant output) or a system property (like stability), which should be achieved by manipulating the outputs not directly but indirectly via the input signals and through the dynamics of the system. *The dynamic nature of the system, which is encoded in the system model, gives a natural constraint on what we can achieve.* This fact is reflected in the control aim sentences above, *i.e.* "as close as possible", "as quick as possible", and so on.

9.1.2 Feedback

The Role of Feedback. Feedback is a central notion in control theory. Most of the controllers, but not all of them, apply either state or output feedback to compute the control action, that is, the input value which is needed to achieve the control aim.

Consider the general input–output relation $y = \mathbf{S}[u]$ introduced in Chapter 2. The most common control objective is to select the input u such that the system output follows a given reference r, *i.e.*

$$r = \mathbf{S}[u] \tag{9.1}$$

This can be solved theoretically if we manage to express u from (9.1). However, generally, we cannot solve this problem. In many cases, it is possible to approximate the solution, but this approach usually does not work well in practice because there are usually unknown *disturbances* that affect the process and the plant dynamics is rarely known completely, *i.e.* there are *model uncertainties*.

As we will see soon, feedback is able to change some of the key dynamic properties of the system completely, while other properties might be invariant to feedback. One reason for applying feedback is that a well-chosen feedback is able to reduce the effect of uncertainty. That is why controllers designed for rough linear models may work reasonably well with the original nonlinear system. On the other hand, feedback is the only tool that stabilizes unstable systems.

9.1.3 Different Kinds of Feedback

In the next few paragraphs, we define the different kinds of feedback that are most often applied to a nonlinear system of the form (3.19), *i.e.*

$$\dot{x}(t) = f(x(t)) + \sum_{i=1}^{m} g_i(x(t))u_i(t)$$
$$y(t) = h(x(t))$$

Static feedbacks make instant relations between the output (or state) and input variables, while *dynamic* feedbacks bring additional dynamics (*i.e.* new state variables) into the control loop. *Output feedbacks* use only output information to generate input, while *state feedbacks* process the whole state vector.

Output feedback has been applied from the very beginning of the existence of controllers. The commonly used PID controller can also be seen as a controller for single-input–single-output systems based on dynamic output feedback with constant parameters. It is important to know, however, that output feedback cannot stabilize even a linear time-invariant system in every case, that is why state feedback is used in advanced control schemes.

As one would naturally expect, generally control goals are easier to achieve when applying state feedback, since the controller can process more information than in the output feedback case. Note that a state feedback is a special case of output feedback, with the output y being the whole state vector x.

Static feedbacks are classified further, according to whether a new reference input is introduced into the closed-loop system or not. The various forms of static feedbacks are defined formally below.

Definition 9.1.1 (Static state feedback)

A static state feedback for the system (3.19) is defined as

$$u = \alpha(x) \tag{9.2}$$

where $\alpha : \ \mathcal{X} \mapsto \mathbb{R}^m$ *is a smooth function.*

Definition 9.1.2 (Static state feedback with new input)

A static state feedback with new input v is of the form

$$u = \alpha(x) + \beta(x)v \tag{9.3}$$

where $\alpha : \ \mathcal{X} \mapsto \mathbb{R}^m$ *and* $\beta : \ \mathbb{X} \mapsto \mathbb{R}^{m \times m}$ *are smooth mappings. Furthermore, $\beta(x)$ is invertible for all x and $v \in \mathbb{R}^m$ is a new vector of control variables.*

Definition 9.1.3 (Static output feedback)

A static output feedback for the system (3.19) is defined as

$$u = \alpha(y) \tag{9.4}$$

where $\alpha : \ \mathbb{R}^p \mapsto \mathbb{R}^m$ *is a smooth mapping.*

Definition 9.1.4 (Static output feedback with new input)

A static output feedback with new input v is a feedback of the form

$$u = \alpha(y) + \beta(y)v \tag{9.5}$$

where $\alpha : \ \mathbb{R}^p \mapsto \mathbb{R}^m$ *and* $\beta : \ \mathbb{R}^p \mapsto \mathbb{R}^{m \times m}$ *are smooth mappings. Furthermore, $\beta(x)$ is invertible for all y and $v \in \mathbb{R}^m$ is a new vector of control variables.*

Figure 9.1 illustrates the definition of static output feedback with new input by a signal flow diagram. Note that Definitions 9.1.1, 9.1.2 and 9.1.3 are all special cases of Definition 9.1.4.

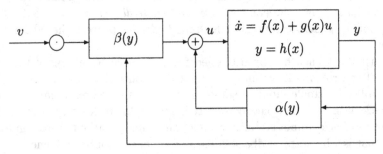

Figure 9.1. Static output feedback with new input

In the case of a dynamic feedback of any type, one usually considers a new input. Therefore the definitions of dynamic state and output feedbacks are given in only one version.

Definition 9.1.5 (Dynamic state feedback)
A dynamic state feedback is defined as

$$
\begin{aligned}
\dot{z} &= \eta(z, x) + \theta(z, x)v \\
u &= \alpha(z, x) + \beta(z, x)v
\end{aligned} \tag{9.6}
$$

where $z \in \mathbb{R}^l$, $\eta : \mathbb{R}^l \times \mathcal{X} \mapsto \mathbb{R}^l$, $\theta : \mathbb{R}^l \times \mathcal{X} \mapsto \mathbb{R}^{l \times m}$, $\alpha : \mathbb{R}^l \times \mathcal{X} \mapsto \mathbb{R}^m$ and $\beta : \mathbb{R}^l \times \mathcal{X} \mapsto \mathbb{R}^{m \times m}$ are smooth mappings and $v \in \mathbb{R}^m$ is the new input vector.

Definition 9.1.6 (Dynamic output feedback)
A dynamic output feedback is defined as

$$
\begin{aligned}
\dot{z} &= \eta(z, y) + \theta(z, y)v \\
u &= \alpha(z, y) + \beta(z, y)v
\end{aligned} \tag{9.7}
$$

where $z \in \mathbb{R}^l$, $\eta : \mathbb{R}^l \times \mathbb{R}^p \mapsto \mathbb{R}^l$, $\theta : \mathbb{R}^l \times \mathbb{R}^p \mapsto \mathbb{R}^{l \times m}$, $\alpha : \mathbb{R}^l \times \mathbb{R}^p \mapsto \mathbb{R}^m$ and $\beta : \mathbb{R}^l \times \mathbb{R}^p \mapsto \mathbb{R}^{m \times m}$ are smooth mappings and $v \in \mathbb{R}^m$ is the new input vector.

Figure 9.2 shows the signal flow diagram of a system controlled by a dynamic output feedback.

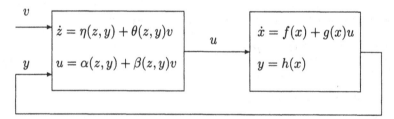

Figure 9.2. Dynamic output feedback

In any type of feedbacks above, we can speak about:

- *Linear feedback*
  When the functions α and β in the feedback equations are linear.
- *Full feedback*
  When every entry of the signal used in the feedback (state or output) is taken into account when computing the input signal.

9.1.4 Linear Static Full State Feedback Applied to SISO LTI Systems

In the example of single-input–single-output (SISO) linear time-invariant (LTI) systems, we show how a full static feedback can be used to change the stability properties of the system.

Let us have a state-space realization (A, B, C) of a single-input–single-output (SISO) LTI system **S**:

$$
\begin{aligned}
&\dot{x}(t) = Ax(t) + Bu(t) \\
&y(t) = Cx(t) \\
&y(t),\ u(t) \in \mathbb{R}, \quad x(t) \in \mathbb{R}^n \\
&A \in \mathbb{R}^{n \times n},\ B \in \mathbb{R}^{n \times 1},\ C \in \mathbb{R}^{1 \times n}
\end{aligned}
\tag{9.8}
$$

with transfer function

$$
H(s) = \frac{b(s)}{a(s)} = \frac{b_1 s^{n-1} + \ldots + b_n}{s^n + a_1 s^{n-1} + \ldots + a_n}
\tag{9.9}
$$

Let us modify **S** by a linear static full state feedback, *i.e.*

$$
\begin{aligned}
&v = u + kx \\
&k = \begin{bmatrix} k_1 & k_2 & \ldots & k_n \end{bmatrix} \\
&k \in \mathbb{R}^{1 \times n} \quad \textit{(row vector)}
\end{aligned}
\tag{9.10}
$$

to get a new desired characteristic polynomial $\alpha(s)$ with

$$
\deg \alpha(s) = n
$$

The state-space description of the *closed-loop system* **S**$_c$ is

$$
\begin{aligned}
&\dot{x}(t) = (A - Bk)x(t) + Bv(t) \\
&y(t) = Cx(t)
\end{aligned}
\tag{9.11}
$$

with the closed-loop polynomial $a_c(s)$

$$
a_c(s) = \det (sI - A + Bk) \quad := \alpha(s)
\tag{9.12}
$$

while the polynomial $a(s)$ for the open-loop system is

$$
a(s) = \det (sI - A)
$$

Observe that the applied linear static full state feedback influences the stability of the system as the state matrix of the closed-loop system has changed to be $A_{cl} = (A - Bk)$.

9.2 Pole-placement Controller for SISO LTI Systems

Pole-placement controllers are the most simple controllers which can be applied only to SISO LTI systems. Although their direct importance of controlling nonlinear systems is not at all great, their design and properties teach us some important lessons.

9.2.1 Problem Statement

Given a finite dimensional LTI system with realization (A, B, C) where the poles of the system are determined by the polynomial $a(s)$. Influence the poles by linear static full state feedback to get a specified (desired) polynomial $\alpha(s)$ such that

$$\deg a(s) = \deg \alpha(s) = n$$

A sub-problem of this problem statement is to find a feedback that *stabilizes* the system.

The pole-placement controller design is a simple but not very advanced design method to achieve the above goal. Notice that there is no explicit loss function expressing the control aim. This means that *this type of controller is not an optimizing one.*

9.2.2 Solution for the SISO Case

We use the following notations:

$$C = [\, B \quad AB \quad A^2B \quad ... \quad A^{n-1}B \,], \quad T_\ell^T = \begin{bmatrix} 1 & a_1 & a_2 & ... & a_{n-1} \\ 0 & 1 & a_1 & ... & a_{n-2} \\ 0 & 0 & 1 & ... & a_{n-3} \\ . & . & . & .... & . \\ 0 & 0 & 0 & ..0 & 1 \end{bmatrix}$$

where C is the controllability matrix and T_ℓ^T is a Toeplitz matrix containing the coefficients a_i of the open-loop characteristic polynomial $\alpha(s)$.

With the above matrices, the state feedback parameters k resulting in a desired characteristic polynomial $\alpha(s)$ can be computed as

$$k = (\underline{\alpha} - \underline{a})T_\ell^{-T}C^{-1} \tag{9.13}$$

where $\underline{\alpha}$ and $\underline{a}$ are vectors composed of the coefficients of the characteristic polynomials $\alpha(s)$ and $\beta(s)$.

As the Toeplitz matrix T_ℓ^T is always nonsingular, the right-hand side of the equation below is nonsingular if and only if C is nonsingular, i.e. (A, B) is a controllable pair.

This shows that a *static full state feedback can arbitrarily relocate the eigenvalues of a system with realization (A, B, C) if and only if (A, B) is controllable.*

It is important to note that the feedback gain k depends on the realization (A, B, C) because the controllability matrix C depends on it.

The software package MATLAB® has a simple procedure call (**place**) to compute the feedback gain vector from the state-space representation matrices of the open-loop system and from the desired poles.

Example 9.2.1 (LTI pole-placement controller)
Pole-placement controller design for a simple LTI system

Consider a simple single-input–single-output LTI system in the form of Equation (9.8) with the matrices

$$A = \begin{bmatrix} 3 & 4 \\ 5 & 6 \end{bmatrix}, \quad B = \begin{bmatrix} 7 \\ 8 \end{bmatrix}, \quad C = [\,1 \ \ 0\,] \qquad (9.14)$$

Let the desired poles of the closed-loop system be -1 and -3. It is easy to check that the eigenvalues of the state matrix above are $\lambda_1 = 7.5$, $\lambda_2 = 2.5$, that is, the system is unstable.
Then the pole-placement controller design procedure checks the full rank of the controllability matrix and gives the following feedback gain vector:

$$k = [\,0.5455 \ \ 1.2727\,] \qquad (9.15)$$

9.3 LQR Applied to LTI Systems

The Linear Quadratic Regulator (LQR) is one of the most commonly used and powerful ways of controlling linear systems. Its extensions are available for stochastic or discrete time systems, as well as for the servo case. Nonlinear systems are also often controlled by LQRs.

9.3.1 Problem Statement

Given a multiple-input–multiple-output (MIMO) LTI system in the following state-space representation form:

$$\begin{aligned} \dot{x}(t) &= Ax(t) + Bu(t), \quad x(0) = x_0 \\ y(t) &= Cx(t) \end{aligned} \qquad (9.16)$$

Define the functional $J(x, u)$ as follows:

$$J(x, u) = \frac{1}{2} \int_0^T [x^T(t) Q x(t) + u^T(t) R u(t)] dt \qquad (9.17)$$

with the *positive semi-definite* state weighting matrix Q and the *positive definite* control weighting matrix R. *Note that the first term in the integrand above measures the derivation of the state from the reference state $x_d = 0$ and the second term measures the control energy.*

It is important to note that both terms in Equation (9.17) are *quadratic forms* in the form of $x^T Q x = \sum_{i,j=1}^{n} x_i Q_{ij} x_j$, which is a scalar-valued positive function for any x if the matrix Q is positive definite.

The *control problem* is to find a control input $\{u(t), \quad t \in [0,T]\}$ that minimizes J subject to Equation (9.16) (constraint on the state variables).

9.3.2 The Solution Method: Calculus of Variations

The solution of the LQR control problem as an optimization problem is based on the *calculus of variations*.

Our previous problem can be generalized to the following mathematical problem:

Minimize

$$J(x, u) = \int_0^T F(x, u, t) dt \tag{9.18}$$

with respect to u subject to $\dot{x} = f(x, u, t)$.

Adjoin the constraint $f(x, u, t) - \dot{x} = 0$ to the cost $J(x, u)$ using a vector Lagrange multiplier $\lambda(.)$ to obtain

$$J(x, \dot{x}, u) = \int_0^T [F(x, u, t) + \lambda^T(t)(f(x, u, t) - \dot{x})] dt \tag{9.19}$$

which will be minimized with respect to $u(t)$. Further define

$$H = F + \lambda^T f$$

as the Hamiltonian function to the optimal control problem.

Integrating Equation (9.19) by part and substituting H, we obtain

$$J = \int_0^T [H + \dot{\lambda}^T x] dt - [\lambda^T x]_0^T$$

If a minimizing u were found, then an arbitrary δu in u should produce $\delta J = 0$:

$$\delta J = -\lambda^T \delta x |_0^T + \int_0^T \left[(\frac{\partial H}{\partial x} + \dot{\lambda}^T) \delta x + \frac{\partial H}{\partial u} \delta u \right] dt$$

The variation δJ above is equal to 0 if and only if the two variations in the integrand above are equal to 0.

Euler-Lagrange Equations. The above condition on minimality of δJ results in the so-called *Euler-Lagrange equations*

$$\frac{\partial H}{\partial x} + \dot{\lambda}^T = 0, \quad \frac{\partial H}{\partial u} = 0 \tag{9.20}$$

The Euler-Lagrange equations for the LTI case are in the following form using

$$\frac{\partial}{\partial x}(x^T Q x) = 2x^T Q$$

$$\dot{\lambda}^T + x^T Q + \lambda^T A = 0, \quad \lambda(T) = 0 \tag{9.21}$$

$$u = -R^{-1} B^T \lambda \tag{9.22}$$

with the system model

$$\dot{x} = Ax(t) + Bu(t), \quad x(0) = x_0 \tag{9.23}$$

From the equations above, we get

$$\begin{bmatrix} \dot{x}(t) \\ \dot{\lambda}(t) \end{bmatrix} = \begin{bmatrix} A & -BR^{-1}B^T \\ -Q & -A^T \end{bmatrix} \begin{bmatrix} x(t) \\ \lambda(t) \end{bmatrix}, \quad \begin{matrix} x(0) = x_0 \\ \lambda(T) = 0 \end{matrix} \tag{9.24}$$

Note that the first equation in the vector differential equation above describes the *system dynamics* and the second one is the so-called *Hammerstein co-state differential equation*.

The above formalism adjoins a *co-state variable* λ *to the original state variable* x for which the following basic lemma holds:

Lemma 9.3.1. *If* (A, B) *is controllable and* (C, A) *is observable, then*

$$\lambda(t) = K(t)x(t) \tag{9.25}$$

where matrix $K(t) \in \mathcal{R}^{n \times n}$

Having the above important lemma, we can substitute Equation (9.25) into Equation (9.21) to obtain

$$\dot{K}x + K\dot{x} = -A^T Kx - Qx$$

If we substitute further Equation (9.23) into $\dot{x}$ in the equation above, we get

$$\dot{K}x + K[A - BR^{-1}B^T K]x + A^T Kx + Qx = 0$$

for any $x(t)$. Therefore the following *Matrix Ricatti Differential Equation* holds for $K(t)$ as an independent variable:

$$\dot{K} + KA + A^T K - KBR^{-1}B^T K + Q = 0 \tag{9.26}$$

Special Case: Stationary Solution. Let us take the special case $T \to \infty$, which is called the stationary solution when

$$J = \int_0^\infty (x^T Q x + u^T R u) dt$$

One can prove that in this case

$$\lim_{t \to \infty} K(t) = K \quad i.e. \quad \dot{K} = 0$$

where K is a constant matrix.

Then Equation (9.26) specializes to the following equation called the *Control Algebraic Ricatti Equation (CARE)*:

$$KA + A^T K - KBR^{-1}B^T K + Q = 0 \tag{9.27}$$

It is easy to see that by taking the transpose of the equation above that K is symmetric.

Let us choose $Q = C^T C$. Then

$$J(x, u) = \int_0^\infty (x^T C^T C x + u^T R u) dt = \int_0^\infty (y^T y + u^T R u) dt \tag{9.28}$$

Theorem 9.3.1. *(Due to R. Kalman). If (C, A) is observable and (A, B) is controllable then CARE has a unique positive definite symmetric solution K.*

9.3.3 LQR as a Full-state Feedback Controller

With this optimal solution, the *optimal feedback is a full state feedback* (from Equations (9.25) and (9.22)) in the form

$$u^0(t) = -R^{-1}B^T K x(t) = -Gx(t) \tag{9.29}$$

where $G = R^{-1}B^T K$.

With the optimal feedback above, *the closed-loop system* equations are as follows:

$$\begin{aligned} \dot{x} &= Ax - BR^{-1}B^T K x = (A - BG)x, \quad x(0) = x_0 \\ y &= Cx \end{aligned} \tag{9.30}$$

Properties of the Closed-loop System. The closed-loop LTI system controlled by an LQR has the following remarkable properties:

- The closed-loop system is asymptotically stable, *i.e.*

$$Re\ \lambda_i(A - BG) < 0, \quad i = 1, 2, ..., n$$

- Stability of the closed-loop system is guaranteed by LQR no matter what the values of A, B, C, R and Q are.

- With A, B given, the specific location of the closed-loop poles depend on the choice of Q and R (*i.e.* Q and R are the design parameters).

The software package MATLAB® has a simple procedure call (**lqr**) to compute the feedback gain vector from the state-space representation matrices of the open-loop system and from the weighting matrices in the loss functional.

Example 9.3.1 (LQR for an LTI system)
LQR design for a simple LTI system introduced in Example 9.2.1

Consider a simple single-input–single-output LTI system in the form of Equation (9.8) with the matrices in Equation (9.14).
Let the weighting matrices in the loss function (9.17) be as follows:

$$Q = \begin{bmatrix} 1 & 0 \\ 0 & 1 \end{bmatrix}, \quad R = \begin{bmatrix} 0.5 \end{bmatrix} \tag{9.31}$$

This means that we weight the error in the state as being much higher than the energy put in for changing the input.
Then the LQR design procedure in MATLAB® gives the following feedback gain vector:

$$k = \begin{bmatrix} 0.9898 & 0.8803 \end{bmatrix} \tag{9.32}$$

The poles of the closed-loop system

$$\begin{bmatrix} -15.2021 & -8.7690 \end{bmatrix}$$

have indeed negative real parts, which means that the closed-loop system is stable.

9.4 Hamiltonian View on Controlled Systems

In this section it is shown how the Hamiltonian view can be applied to deriving controllers by constructing a Hamiltonian description of controlled systems.

The direct relationship with classical mechanics described in Section 8.3 will be used here in the context of LTI systems where a quadratic Lyapunov function can always be constructed (see Theorem 7.3.3 in Chapter 7). There is also a relationship with the well-known formulation of designing LQR controllers for LTI systems.

Hamiltonian Formulation of Input Design. Let us choose the Hamiltonian function of a concentrated parameter system in the following form:

$$H(x, u, \lambda, t) \overset{\Delta}{=} F(x, u, t) + \lambda^T(t)f(x, u, t), \qquad \lambda(t) \overset{\Delta}{=} p(t) \tag{9.33}$$

where the state equation of the system is

$$\dot{x} = f(x, u, t), \qquad x(0) = x_0 \tag{9.34}$$

We aim at finding an input u such that the functional

$$J(x, u) = \int_0^T F(x, u, t)dt \rightarrow \min_{u \in \mathcal{L}_2[0,T]}. \tag{9.35}$$

is minimized. Here $F(x, u, t)$ is a given optimality criterion, the analog for potential energy, while $\lambda^T(t)f(x, u, t)$ corresponds to the kinetic energy.

Let us investigate the relationship of the Hamiltonian K_γ in Equation (8.8) of Section 8.3 derived from the storage function and the definition of H in Equation (9.33) with the special case of input-affine nonlinear state-space models, when $f(x, u, t) = f(x) + g(x)u$. Comparing the equations, we find that

$$\lambda = p = V_x^T \tag{9.36}$$

$$F(x, u, t) = -\frac{1}{2}[\gamma^2 \|u\|^2 - h^T(x)h(x)] \tag{9.37}$$

Observe that $F(x, u, t)$ is the supply rate and the co-state variable plays the same role as before, joining the right-hand side of the state equation to the Hamiltonian.

From the above, it follows that *if we manage to choose the optimality criterion for a controller such that a storage function can be derived from it in the same way as (9.37), then the controller will stabilize the original system.*

The LTI Case. This is exactly the case of LQR for LTI systems as was seen in Section 9.3 before. In this case

$$F(x, u, t) = x^T Q x + u^T R u$$

which is in good agreement with the open-loop natural Lyapunov (storage) function

$$V(x) = x^T P x$$

With the Hamiltonian view on controlled systems, we can derive the so-called Hamiltonian equations in the following way. We adjoin the constraint $f(x, u, t) - \dot{x} = 0$ to the cost $J(x, u)$ by using a vector Lagrange multiplier $\lambda(.)$ to obtain

$$J(x, \dot{x}, u) = \int_0^T [F(x, u, t) + \lambda^T(t)(f(x, u, t) - \dot{x})]dt \qquad (9.38)$$

which will be minimized with respect to $u(t)$. Integrating Equation (9.38) by part and substituting H, we obtain

$$J = \int_0^T [H + \dot{\lambda}^T x]dt - [\lambda^T x]_0^T$$

If a minimizing u were found, then an arbitrary δu in u should produce $\delta J = 0$:

$$\delta J = -\lambda^T \delta x|_0^T + \int_0^T \left[(\frac{\partial H}{\partial x} + \dot{\lambda}^T)\delta x + \frac{\partial H}{\partial u}\delta u \right] dt$$

The variation δJ above is equal to 0 if and only if the two variations in the integrand above are equal to 0.

The Hamiltonian equations relate the state and co-state variables as follows:

$$\dot{x} = \frac{\partial H(x, p, u)}{\partial p} \qquad (9.39)$$

$$\dot{p} = -\frac{\partial H(x, p, u)}{\partial x} \qquad (9.40)$$

where $\dot{\lambda} = \dot{p}$.

The necessary condition for the optimality of control $u^*(t)$, $t \in [0, T]$ is:

$$\frac{\partial H(x, p, u)}{\partial u} = 0 \qquad (9.41)$$

The Euler–Lagrange equations can be derived from (9.40) and (9.41) and from the defining Equation (9.33):

$$\frac{\partial F}{\partial x} + \frac{\partial f^T}{\partial x}\lambda + \dot{\lambda} = 0 \qquad (9.42)$$

$$\frac{\partial F}{\partial u} + \frac{\partial f^T}{\partial u}\lambda = 0 \qquad (9.43)$$

It is important to note that now we have arrived at the same equations as before in Equations (9.20) in Section 9.3 where they have been obtained by solving a control optimization problem using calculus of variations.

The solution of the Euler-Lagrange equations above is performed for LTI systems in Section 9.3 before assuming that

$$\lambda(t) = K(t)x(t) \qquad (9.44)$$

where K is a suitable time-dependent square and symmetric matrix.

9.5 Case Study: Linear Full-state Feedback Control of a Continuous Fermenter

The usual way of approaching a nonlinear system to be controlled is to start with a robust linear full-state feedback controller which is designed by using a linearized model of the system around a nominal operating point. Then the stability region of the closed-loop controlled system can be estimated by using quadratic Lyapunov functions (see Subsection 7.3.2 for details).

The aim of this section is to show the possible difficulties which can be met using this linearized approach [76].

The simple continuous fermenter is used here for this purpose. The input-affine nonlinear state-space model of the fermenter is developed in Subsection 4.5.3 and is given in Equations (4.78)–(4.79). The variables and parameters of the fermentation process model are collected in Table 4.1.

The linearized state-space model of the continuous fermenter is also developed in Equations (4.84)–(4.87).

9.5.1 Pole-placement Controller

The purpose of this section is:

- to provide a simple controller design approach for later comparison,
- to examine the possibilities of stabilizing the system by partial feedback (preferably by feeding back the substrate concentration only).

Pole-placement by Full State Feedback. First, a full state feedback is designed such that the poles of the linearized model of the closed-loop system are at $[-1 \quad -1.5]^T$. The necessary full state feedback gain is

$$K_{pp} = [-0.3747 \quad 0.3429] \tag{9.45}$$

A simulation run is shown in Figure 9.3 starting from the initial state $X(0) = 0.1\frac{g}{l}$ and $S(0) = 0.5\frac{g}{l}$. It is clearly seen that the closed-loop nonlinear system has an additional undesirable stable equilibrium point and the controller drives and stabilizes the system towards this point. This stable undesired operating point can be easily calculated from the state equations and the parameters of the closed-loop system: $X = 3.2152\frac{g}{l}$, $S = 3.5696\frac{g}{l}$. The time derivative of the Lyapunov function is shown in Figure 9.4. The appearance of the undesired stable operating point warns us *not to apply linear controllers based on locally linear models for a nonlinear system without careful prior investigation.*

Partial Linear Feedback. Motivated by the fact that the zero dynamics of the fermenter (see Subsection 5.5.2 for details) is globally asymptotically stable when the substrate concentration is the output, let us consider the following static partial state feedback:

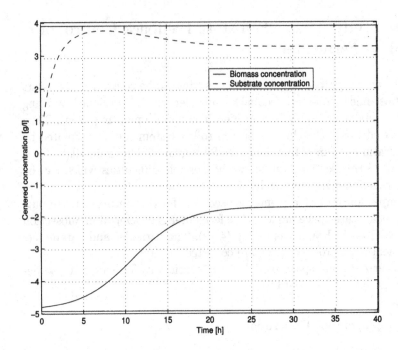

Figure 9.3. Centered state variables and input, full state feedback pole-placement controller, $X(0) = 0.1\frac{g}{l}$, $S(0) = 0.5\frac{g}{l}$

$$u = Kx$$

where $K = [0 \;\; k]$, *i.e.* we only use the substrate concentration for feedback.

The *stability region of the closed-loop system* is investigated by using the time derivative of the quadratic Lyapunov function. Figure 9.5 shows that the stability region of the closed-loop system is quite wide. Furthermore, it can be easily shown that for, *e.g.* $k = 1$, the only stable equilibrium point of the closed-loop system (except for the wash-out steady state) is the desired operating point. The eigenvalues of the closed-loop system with $k = 1$ are -0.9741 and -2.6746.

9.5.2 LQ Control

Usually the well-known LQ-controller is used as a reference controller for comparison and tuning the nonlinear controllers. LQ-controllers are popular and widely used for process systems. They are known to stabilize any stabilizable linear time-invariant system globally, that is, over the entire state-space. This type of controller is designed for the locally linearized model of the process and minimizes the cost function

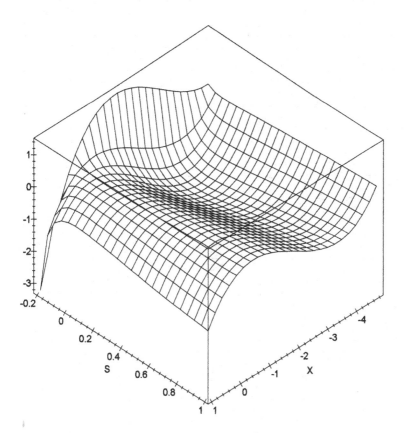

Figure 9.4. Time derivative of the Lyapunov function as a function of centered state variables $q_1 = 1, q_2 = 0.1$, pole-placement controller, $K_{pp} = [-0.3747 \; 0.3429]$

$$J(x(t), u(t)) = \int_0^\infty \left(x^T(t) R_x x(t) + u^T(t) R_u u(t) \right) dt \qquad (9.46)$$

where R_x and R_u (the design parameters) are positive definite weighting matrices of appropriate dimensions. The optimal input that minimizes the above functional is in the form of a linear full state feedback controller $u = -Kx$. The results for two different weighting matrix selections are investigated.

Cheap Control. In this case, the design parameters R_x and R_u are selected to be $R_x = I^{2 \times 2}$ and $R_u = 1$. The resulting full state feedback gain is $K = [-0.6549 \; 0.5899]$.

Expensive Control. The weighting matrices in this case are $R_x = 10 \cdot I^{2 \times 2}$ and $R_u = 1$. There are no significant differences in terms of controller performance compared to the previous case. The full state feedback gain in this case is $K = [-1.5635 \; 2.5571]$.

The stability region is again investigated using the time derivative of a quadratic Lyapunov function. The time-derivative function as a function

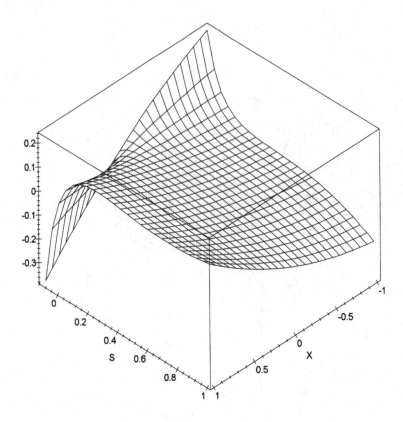

Figure 9.5. Time derivative of the Lyapunov function as a function of centered state variables $q_1 = 1, q_2 = 0.1$, partial linear feedback, $k = 1$

of the centered state variables for cheap and expensive control is seen in Figures 9.6 and 9.7 respectively. Unlike the linear case where LQR always stabilizes the system, it is seen that the stability region does *not* cover the entire operating region. Indeed, a simulation run in Figure 9.8 starting with an "unfortunate" initial state exhibits unstable behavior for the nonlinear fermenter.

Note that an LQ-controller is structurally the same as a linear pole-placement controller (*i.e.* a static linear full state feedback). Therefore *undesired stable steady states may also appear depending on the LQ-design.*

9.6 Further Reading

[64] discusses multivariable controller design methods for linear systems with several process system case studies. One of the most popular books of the recent decade on linear controller design is [48]. The first papers of great

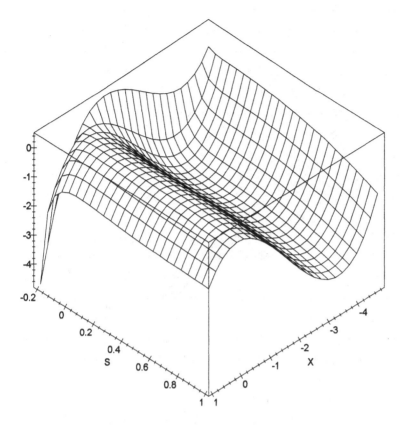

Figure 9.6. Time derivative of the Lyapunov function as a function of centered state variables $q_1 = 1, q_2 = 1$, LQ-controller, cheap control

influence on optimal control were [40] and [41]. The LQR problem is covered in great detail in, *e.g.* [6].

Classical control-related problems for LPV systems are solved in [78], [74] and [69] using a geometric approach.

9.7 Summary

The basic notions of controller design, including feedback of different types, control design by optimization and the Hamiltonian view on controlled systems, are described in this chapter.

The most simple linear controllers, the pole-placement controller and the linear quadratic regulator (LQR) are also introduced here as applied to LTI systems.

A case study of a simple nonlinear continuous fermenter model is used to illustrate how to design and use linear controllers to nonlinear process

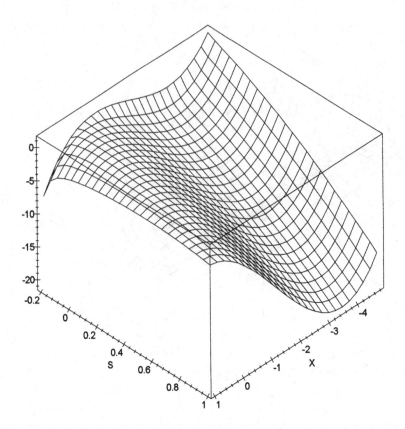

Figure 9.7. Time derivative of the Lyapunov function as a function of centered state variables $q_1 = 1, q_2 = 1$, LQ controller, expensive control

systems and what the dangers are of applying them without careful prior investigation.

9.8 Questions and Exercises

Exercise 9.8.1. Compare the pole-placement and LQR design for LTI systems using Examples 9.2.1 and 9.3.1. Comment on the similarities and differences.

Exercise 9.8.2. Classify controllers according to their control aim. Give simple example(s) to each of your class.

Exercise 9.8.3. Give the *joint* necessary condition of the applicability of pole-placement controllers and LQR in the case of LTI systems.

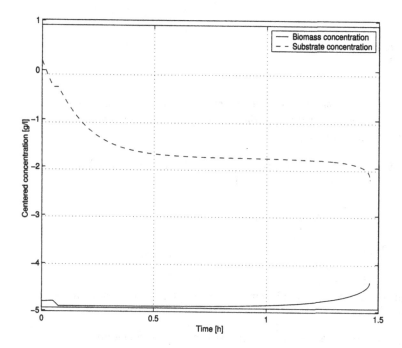

Figure 9.8. Unstable simulation run, LQ-controller, expensive control, $X(0) = 0.1\frac{g}{l}$, $S(0) = 0.5\frac{g}{l}$

Exercise 9.8.4. Explain the significance and use of the Hamiltonian description in controller design for both the LTI and the nonlinear input-affine case.

Exercise 9.8.5. The state-space model of a linear system is given by the following matrices:

$$A = \begin{bmatrix} 1 & 0 \\ 0 & 2 \end{bmatrix} \quad B = \begin{bmatrix} 1 \\ 4 \end{bmatrix} \quad C = \begin{bmatrix} 1 & 0 \\ 0 & 1 \end{bmatrix} \tag{9.47}$$

Calculate the state feedback gain such that the poles of the closed-loop system are at $[-1 \ -2]^T$.

Exercise 9.8.6. Consider a first-order process with the state-space realization

$$\dot{x}(t) = x(t) + u(t), \quad y(t) = x(t)$$

Calculate the optimal state feedback gain that minimizes the functional

$$J(x(t), u(t)) = \int_0^\infty x(t)^T Q x(t) + u(t)^T R u(t) dt \tag{9.49}$$

for the following cases:

- $Q = 1$, $R = 1$
- $Q = 1$, $R = 10$
- $Q = 1$, $R = 0.1$
- $Q = 10$, $R = 1$

Examine the poles of the closed-loop systems for the above four cases. Which control is the cheapest if we consider the cost of input energy? Which one provides the quickest response?

Exercise 9.8.7. Consider the state equation of the simple quasi-polynomial system in Exercise 6.9.4 in the form of

$$\frac{dx_1}{dt} = x_1^{1/2} x_2^{3/2} - x_1 x_2^3 + x_1^{1/2} u_1$$

$$\frac{dx_2}{dt} = 3x_1 x_2^{1/2} - 2x_1^2 x_2$$

with the output equation $y = x_1$.

Design a stabilizing pole-placement controller for the system in the following steps:

1. Apply the nonlinear coordinate transformation (see Definition 3.5.2)

$$\Phi(x) = \begin{bmatrix} \ln(x_1) \\ \ln(x_2) \end{bmatrix}$$

   to the state equation to get a LTI model in the new coordinates.
2. Design a pole-placement controller in the new coordinate system moving the poles to $[-1, \; -1]$.
3. Express the resulted feedback in the original coordinates.

Characterize the feedback you have obtained.

Exercise 9.8.8. What is a joint necessary condition of the applicability of pole-placement control and LQR in the case of LTI systems?

10. Feedback and Input–output Linearization of Nonlinear Systems

The aim of linearization is to apply a suitable nonlinear coordinate transformation to a nonlinear system in order to obtain a linear one in the new co-ordinates. As will be shown in this chapter, the coordinates transformation must be supplemented by a static nonlinear feedback to achieve linearization. Therefore it is a basic but limited technique for control of nonlinear systems.

The description in this chapter mainly follows the well-known book of Isidori [37]. The material is broken down into the following sections:

- *Relative degree*
  We start with the notion and computation of the relative degree of a SISO nonlinear system, which is one of the basic properties of a nonlinear system.
- *Exact linearization via state feedback*
  The exact linearization procedure is described in detail and illustrated with examples.
- *Input–output linearization*
  Input–output linearization is applied when exact linearization is not applicable or not feasible, therefore it is of great practical importance.
- *Process systems with maximal relative degree*
  The engineering conditions of process systems, being of maximal relative degree, are investigated in this section. This property is required for the application of the basic case of feedback linearization.
- *Case study: feedback and input–output linearization of a continuous fermenter*
  The controller design techniques based on feedback and input–output linearization of a simple continuous fermentation process are described and compared here.
- *Output selection for feedback linearization*
  This section deals with a controller structure selection problem on how to select a suitable output from the possible ones for feedback linearization.

10.1 Relative Degree

The notion of relative degree plays a central role not only in the necessary condition for performing exact linearization but also in other parts of nonlinear control theory. This is the reason why the notion of relative degree has been introduced earlier by Definition 5.5.2 in Section 5.5, which is repeated here for convenience.

The single-input–single-output nonlinear system

$$\dot{x} = f(x) + g(x)u$$
$$y = h(x)$$

is said to have relative degree r at a point x^0 if

1. $L_g L_f^k h(x) = 0$ for all x in a neighborhood of x^0 and all $k < r - 1$.
2. $L_g L_f^{r-1} h(x^0) \neq 0$.

Here it is important to remember that *the relative degree r is exactly equal to the number of times one has to differentiate the output $y(t)$ at time $t = t^0$ in order to have the value $u(t^0)$ explicitly appearing.*

To show this, we first observe that the function $g(x)$ in Equation (5.51) (repeated above) is vector-valued, that is

$$g(x) = \begin{bmatrix} g_1(x) \\ .. \\ g_n(x) \end{bmatrix}$$

Then we compute the time derivative of the output equation several (k) times as follows:

- $k = 0$: $y = h(x)$

$$\frac{dy}{dt} = \frac{\partial h}{\partial x}\dot{x} = \frac{\partial h}{\partial x}(f(x) + g(x)u) = L_f h + L_g h \cdot u$$

- $k = 1$:   assume $L_g h = 0$ (condition 1) then $\dot{y} = L_f h$, and

$$\frac{d^2 y}{dt^2} = \frac{\partial (L_f h)}{\partial x}(f(x) + g(x)u) = L_f^2 h + L_g L_f h \cdot u$$

If $L_g L_f h \neq 0$ then $r = 2$ (see condition 2), otherwise we continue the procedure, and so on.

Note that *for MIMO systems the relative degree can be defined pairwise for all of the possible input–output pairs.*

The following simple example shows how the concept of relative degree applies for SISO LTI systems.

Example 10.1.1 (Relative degree of SISO LTI systems)
Relative Degree of Linear Time-invariant Systems

Let us consider the following SISO LTI system:

$$\dot{x} = Ax + Bu \tag{10.1}$$
$$y = Cx \tag{10.2}$$

It is easy to see that

$$L_f^k h(x) = CA^k x \tag{10.3}$$

and

$$L_g L_f^k h(x) = CA^k B \tag{10.4}$$

Notice that the relative degree conditions depend on the Markov parameters of an LTI SISO system.
Thus the relative degree r is given by

$$CA^k B = 0 \qquad \forall k < r - 1 \tag{10.5}$$
$$CA^{r-1} B \neq 0 \tag{10.6}$$

It is well known from linear systems theory that the integer satisfying these conditions is exactly equal to the difference between the degree of the denominator polynomial (n) and the degree of the numerator polynomial (m) of the transfer function $H(s) = C(sI - A)^{-1}B$, i.e. $r = n - m \geq 0$.

The next example contains the calculation of relative degree for a simple nonlinear system.

Example 10.1.2 (Relative degree of a nonlinear system)

Calculate the relative degree of the following system at the point $[1\ 1]^T$.

$$\dot{x} = \begin{bmatrix} x_2 \\ 2\omega\zeta(1 - \mu x_1^2)x_2 - \omega^2 x_1 \end{bmatrix} + \begin{bmatrix} 0 \\ 1 \end{bmatrix} u \tag{10.7}$$
$$y = h(x) = x_1 \tag{10.8}$$

where ω, ζ and μ are non-zero constants. If possible, find two points in the state-space where the relative degree of the system is different.

We compute the necessary Lie-derivatives as follows:

$$L_g h(x) = \frac{\partial h}{\partial x} g(x) = \begin{bmatrix} 1 & 0 \end{bmatrix} \begin{bmatrix} 0 \\ 1 \end{bmatrix} = 0 \qquad (10.9)$$

$$L_f h(x) = x_2 \qquad (10.10)$$

$$L_g L_f h(x) = \frac{\partial L_f h}{\partial x} g(x) = \begin{bmatrix} 0 & 1 \end{bmatrix} \begin{bmatrix} 0 \\ 1 \end{bmatrix} = 1 \qquad (10.11)$$

This means that the relative degree of the system is 2 in any point of the state-space (including $[1\ 1]^T$). Therefore there are no such points in the state-space where the relative degree of the system is different.

10.2 Exact Linearization via State Feedback

The aim of exact linearization is to apply a suitable nonlinear static state feedback to a SISO nonlinear system in order to obtain a linear one in the new coordinates and between the original output and the newly introduced transformed input. Thereafter any controller design method, such as PID, pole-placement or LQR, applicable for SISO LTI systems can be used to stabilize the system or modify its dynamic properties (see Chapter 9 for state feedback controllers for LTI systems).

Exact linearization via state feedback is a basic but limited technique for control of nonlinear systems because it is only applicable for systems satisfying a relative degree condition.

10.2.1 Nonlinear Coordinates Transformation and State Feedback

In a SISO system the most convenient structure for a static state feedback control is when the input variable u is computed as

$$u = \alpha(x) + \beta(x)v \qquad (10.12)$$

where v is the external reference input. The composition of this control with a SISO nonlinear input-affine system of the form (5.51)–(5.52) gives

$$\dot{x} = f(x) + g(x)\alpha(x) + g(x)\beta(x)v \qquad (10.13)$$
$$y = h(x) \qquad (10.14)$$

The application that will be discussed is the use of state feedback (and change of coordinates in the state-space) with the purpose of *transforming a given nonlinear system into a linear and controllable one.*

Note that the above nonlinear state feedback is a direct extension of the linear state feedback (see in Section 9.1.2) applied in the problem statement of pole-placement controllers described in Section 9.2:

$$u = -kx + v \quad \text{(LTI)}$$
$$u = \alpha(x) + \beta(x)v \quad \text{(nonlin)}$$

The following lemma introduces a coordinate transformation that is applied in bringing the system into a linear form in the new coordinates.

Lemma 10.2.1. *Consider a nonlinear system having relative degree $r = n$ (i.e. exactly equal to the dimension of the state-space) at some point $x = x^0$ and the following coordinates transformation:*

$$
\Phi(x) = \begin{bmatrix} \phi_1(x) \\ \phi_2(x) \\ \cdots \\ \phi_n(x) \end{bmatrix} = \begin{bmatrix} h(x) \\ L_f h(x) \\ \cdots \\ L_f^{n-1} h(x) \end{bmatrix}
\tag{10.15}
$$

i.e. a transformation by the output function and its first $(n-1)$ derivatives along $f(x)$. Then there exists a nonlinear static feedback in the form of (10.12) such that the closed-loop system in the new coordinates is linear and controllable.

In the new coordinates

$$z_i = \phi_i(x) = L_f^{i-1} h(x), \quad 1 \le i \le n \tag{10.16}$$

the system is given by

$$\dot{z}_1 = z_2 \tag{10.17}$$
$$\dot{z}_2 = z_3 \tag{10.18}$$
$$\cdots \tag{10.19}$$
$$\dot{z}_{n-1} = z_n \tag{10.20}$$
$$\dot{z}_n = b(z) + a(z)u \tag{10.21}$$

since

$$
\frac{d}{dt} L_f^k h(x) = \frac{\partial L_f^k h(x)}{\partial x} \dot{x} = \frac{\partial L_f^k h(x)}{\partial x} (f(x) + g(x)u)
$$
$$
= L_f^{k+1} h(x) + L_g L_f^k h(x) u \tag{10.22}
$$

As the relative degree of the system is equal to n

$$L_g L_f^k h(x) = 0, \quad \forall k < n$$

the above Equation (10.22) specializes to

$$\frac{d}{dt}L_f^k h(x) = \dot{z}_k = L_f^{k+1}h(x) = z_{k+1}, \quad \forall k < n \tag{10.23}$$

and the last equation is in the form

$$\dot{z}_n = \frac{d}{dt}L_f^{n-1}h(x) + L_g L_f^{n-1}h(x) \cdot u \tag{10.24}$$

Suppose that the following state feedback control law is chosen:

$$u = \frac{1}{a(z)}(-b(z) + v) \tag{10.25}$$

with

$$\alpha(x) = -\frac{b(z)}{a(z)} = -\frac{L_f^n h(x)}{L_g L_f^{n-1}h(x)}, \quad \beta(x) = \frac{1}{a(z)} = \frac{1}{L_g L_f^{n-1}h(x)}$$

as follows from Equation (10.24) above.

The resulting closed-loop system is then governed by the equations

$$\dot{z}_1 = z_2 \tag{10.26}$$
$$\dot{z}_2 = z_3 \tag{10.27}$$
$$\cdots \tag{10.28}$$
$$\dot{z}_{n-1} = z_n \tag{10.29}$$
$$\dot{z}_n = v \tag{10.30}$$

i.e. the closed-loop system is indeed linear and controllable.

It is important that the transformation consists of two basic ingredients:

1. A change of coordinates, defined locally around the point x^0.
2. A state feedback, also defined locally around the point x^0.

10.2.2 The State-space Exact Linearization Problem for SISO Systems

A critical assumption in Lemma 10.2.1 is to have the relative degree of the system to be equal to the number of state variables, i.e. $r = n$. If this is not the case, i.e. when $r < n$, one may try to construct an *artificial output* $y = \lambda(x)$ (with $\lambda(x)$ being a nonlinear function different from the original $h(x)$) such that the relative system is of $r = n$. The idea is to use the basic case given in Lemma 10.2.1 to find conditions for the existence and a way to construct $\lambda(x)$.

It is important to note, however, that $\lambda(x)$ is not expected to be unique (if it exists at all) because the nonlinear coordinates transformation $\Phi(x)$ is not unique either. Therefore we would like to find a transformation $\Phi(x)$ which is simple and invertible.

Problem Statement. Consider an input-affine nonlinear system model without the output equation

$$\dot{x} = f(x) + g(x)u \tag{10.31}$$

and suppose the following problem is to be solved. Given a point x^0 in the state-space, find (if possible) a neighborhood U of x^0, a feedback

$$u = \alpha(x) + \beta(x)v \tag{10.32}$$

defined on U, and a coordinates transformation $z = \Phi(x)$ also defined on U, such that the corresponding closed-loop system

$$\dot{x} = f(x) + g(x)\alpha(x) + g(x)\beta(x)v \tag{10.33}$$

in the new coordinates $z = \Phi(x)$ is linear and controllable, $i.e.$

$$\left[\frac{\partial \Phi}{\partial x} (f(x) + g(x)\alpha(x)) \right]_{x = \Phi^{-1}(z)} = Az \tag{10.34}$$

$$\left[\frac{\partial \Phi}{\partial x} (g(x)\beta(x)) \right]_{x = \Phi^{-1}(z)} = B \tag{10.35}$$

for some suitable matrix $A \in \mathbb{R}^{n \times n}$ and vector $B \in \mathbb{R}^n$ satisfying the condition

$$\text{rank}(B \ AB \ \dots \ A^{n-1}B) = n \tag{10.36}$$

This problem is the SISO version of the so-called *State-space Exact Linearization Problem*.

Existence of a Solution. The following lemma gives a necessary and sufficient condition for the solvability of the above State-space Exact Linearization Problem:

Lemma 10.2.2. *The State-space Exact Linearization Problem is solvable if and only if there exists a neighborhood U of x^0 and a real-valued function λ, defined on U, such that the system*

$$\dot{x} = f(x) + g(x)u \tag{10.37}$$
$$y = \lambda(x) \tag{10.38}$$

has relative degree n at x^0.

The problem of finding a function λ such that the relative degree of the system at x^0 is exactly n, $i.e.$ a function such that

$$L_g\lambda(x) = L_gL_f\lambda(x) = \cdots = L_gL_f^{n-2}\lambda(x) = 0 \text{ for all } x \tag{10.39}$$

$$L_gL_f^{n-1}\lambda(x^0) \neq 0 \tag{10.40}$$

is apparently a problem involving the solution of a system of partial differential equations (PDEs).

In order to see this, we rewrite the first equation in (10.39) into PDE form using the fact that $\lambda(x) = \lambda(x_1, \ldots, x_n)$ is a scalar-valued function:

$$L_g\lambda(x) = \sum_{i=1}^{n} \frac{\partial \lambda}{\partial x_i}(x) \cdot g_i(x) = 0$$

with $g_i(x)$ being a given function.

For the purpose of the solution we introduce a short notation

$$L_{ad_fg}\lambda = \frac{\partial \lambda}{\partial x} \cdot [f, g] \tag{10.41}$$

where $ad_f^i g$ is recursively defined as follows:

$$ad_f^0 g(x) = g(x) \tag{10.42}$$
$$ad_f^i g(x) = [f, ad_f^{i-1} g(x)] \tag{10.43}$$

For example

$$ad_f^1 g(x) = ad_f g = [f, g], \quad ad_f^2 g(x) = [f, [f, g]]$$

As a matter of fact, Equations (10.39) are then equivalent to

$$L_g\lambda(x) = L_{ad_fg}\lambda(x) = \cdots = L_{ad_f^{n-2}g}\lambda(x) = 0 \tag{10.44}$$

and the non-triviality condition (10.40) is equivalent to

$$L_{ad_f^{n-1}g}\lambda(x^0) \neq 0 \tag{10.45}$$

By this, we have arrived at a theorem that states the necessary and sufficient conditions for the solvability of the State-space Exact Linearization Problem:

Theorem 10.2.1. *Suppose a system*

$$\dot{x} = f(x) + g(x)u \tag{10.46}$$

is given. The State-space Exact Linearization Problem is solvable near a point x^0 (i.e. there exists an "output" function λ for which the system has relative degree n at x^0) if and only if the following conditions are satisfied:

1. *The matrix* $\Delta_c(x^0) = \left[g(x^0) \; ad_f g(x^0) \; \cdots \; ad_f^{n-2} g(x^0) \; ad_f^{n-1} g(x^0) \right]$ *has rank n.*
2. *The distribution* $D = \text{span}\{g, ad_f g, \ldots, ad_f^{n-2}\}$ *is involutive near x^0.*

The distribution D in condition 2 can also be written in the form

$$D = \text{span}\{ g, [f, g], [f, [f, g]], \ldots \}$$

and its dimension is at most $n - 1$, i.e. $\dim D \leq n - 1$. The difficulty in constructing D lies in the fact that it is difficult to have it involutive.

Solution Procedure. On the basis of the previous discussion, the procedure leading to the construction of a feedback $u = \alpha(x) + \beta(x)v$ and of a coordinates transformation $z = \Phi(x)$ by solving the State-space Exact Linearization problem consists of the following steps:

1. From $f(x)$ and $g(x)$, construct the vector fields
$$g(x), ad_f g(x), \ldots, ad_f^{n-2} g(x), ad_f^{n-1} g(x)$$
and check conditions 1 and 2.
2. If both are satisfied, solve the partial differential equation (10.39) for $\lambda(x)$.
3. Calculate the feedback functions
$$\alpha(x) = \frac{-L_f^n \lambda(x)}{L_g L_f^{n-1} \lambda(x)}, \quad \beta(x) = \frac{1}{L_g L_f^{n-1} \lambda(x)} \tag{10.47}$$

4. Calculate the coordinate transformation
$$\Phi(x) = \mathrm{col}(\lambda(x), L_f \lambda(x), \ldots, L_f^{n-1} \lambda(x)) \tag{10.48}$$

Definition 10.2.1 (Linearizing feedback and coordinates)
The feedback defined by the functions (10.47) is called the linearizing feedback and the new coordinates given by (10.48) are called the linearizing coordinates.

It is important to note that the feedback linearization of SISO nonlinear systems is not necessarily robust (sometimes even highly sensitive) with respect to the parameter or structure mismatches, therefore it is of limited use in process system applications.

10.2.3 Simple Examples for Feedback Linearization

Two simple examples shows the difficulties of feedback linearization using the solution procedure of the State-space Exact Linearization Problem.

Linearization of a One-dimensional System

Problem Statement. Consider a nonlinear system that is given by the following state equation
$$\frac{dx}{dt} = -kx^2 + xu \tag{10.49}$$

where k is a non-zero constant. We aim to find a linearizing feedback and a linearizing coordinate transformation for this system.

The solution proceeds by answering the following questions and solving the simple sub-problems as detailed below.

1. Is it possible to solve the State-space Exact Linearization Problem at any point of the state-space?
2. Choose an appropriate point in the state-space and find a linearizing feedback and linearizing coordinate transformation for the system. Write the equations of the closed-loop system in the original coordinates and also in the transformed coordinates. Examine the stability, controllability and observability of the closed-loop system.
3. Is the above solution unique? If not, try to give another solution.

Solution. We solve the problem in the order of the steps above as follows:

1. If the function λ is chosen, for example, as

$$\lambda(x) = x \qquad (10.50)$$

then the State-space Exact Linearization Problem can be solved around any point of the state-space except 0, since

$$L_g\lambda(x) = L_g L_f^0 \lambda(x) = x \qquad (10.51)$$
$$L_f\lambda(x) = -kx^2 \qquad (10.52)$$
$$\qquad (10.53)$$

2. Thus the linearizing feedback around any point (except 0) is given by

$$\alpha(x) = \frac{kx^2}{x} = kx \qquad (10.54)$$

$$\beta(x) = \frac{1}{x} \qquad (10.55)$$

and the linearizing coordinate transformation is

$$z = \phi(x) = x \qquad (10.56)$$

Since the above coordinate transformation is identical, the state equation of the closed-loop system is the same in the original and the transformed coordinate is in the form

$$\dot{x} = -kx^2 + x(kx + \frac{1}{x}v) = v \qquad (10.57)$$

The closed-loop system is on the boundary of stability, and it is controllable and observable.

3. The above solution is not unique, we can find infinitely many suitable transformations, *e.g.*

$$\lambda(x) = x^2 + 1 \qquad (10.58)$$
$$L_g\lambda(x) = 2x^2 \qquad (10.59)$$
$$L_f\lambda(x) = -2kx^3 \qquad (10.60)$$

$$\alpha(x) = \frac{2kx^3}{2x^2} = kx \qquad (10.61)$$

$$\beta(x) = \frac{1}{2x^2} \qquad (10.62)$$

Then the equations of the closed-loop system are

$$\dot{x} = -kx^2 + x(kx + \frac{1}{2x^2}v) = \frac{1}{2x}v \tag{10.63}$$

$$z = \phi(x) = x^2 + 1 \tag{10.64}$$

$$\dot{z} = 2x\dot{x} = v \tag{10.65}$$

Exact Linearization of a Two-dimensional System.

Problem statement. The aim in this example is to exactly linearize the following simple two-dimensional system around its equilibrium point $x_0 = 0$:

$$\begin{bmatrix} \dot{x}_1 \\ \dot{x}_2 \end{bmatrix} = \begin{bmatrix} -x_1^3 + x_2 \\ -x_1 x_2 \end{bmatrix} + \begin{bmatrix} 0 \\ 1 \end{bmatrix} u \tag{10.66}$$

Solution. First, let us find a function λ for which the system has relative degree 2 at x_0. For this, we would need $L_g\lambda(x) = 0$ in a neighbourhood of x_0, and $L_g L_f \lambda(x_0) \neq 0$. It's easy to see from $g(x)$ in the model that a function λ depending only on x_1 meets the first requirement. Let us select λ as

$$\lambda(x) = x_1$$

and thus $L_g\lambda(x) = [1 \ 0][0 \ 1]^T = 0$. Now, calculate $L_g L_f \lambda$, which is written as

$$L_g L_f \lambda(x) = [-3x_1^2 \ 1][0 \ 1]^T = 1$$

and which satisfies the second condition of having relative degree 2 at any point of the state-space. The linearizing coordinates transformation Φ and its inverse are given by

$$\begin{bmatrix} z_1 \\ z_2 \end{bmatrix} = \Phi(x) = \begin{bmatrix} x_1 \\ -x_1^3 + x_2 \end{bmatrix}$$

and

$$\begin{bmatrix} x_1 \\ x_2 \end{bmatrix} = \Phi^{-1}(z) = \begin{bmatrix} z_1 \\ z_2 + z_1^3 \end{bmatrix}$$

The functions in the nonlinear linearizing feedback can be calculated as

$$\alpha(x) = \frac{-L_f^2 \lambda(x)}{L_g L_f \lambda(x)} = 3x_1^2(x_1^3 - x_2) - x_1 x_2$$

and

$$\beta(x) = \frac{1}{L_g L_f \lambda(x)} = 1$$

The system with the feedback above in the original coordinates is

$$\dot{x}_1 = -x_1^2 + x_2$$
$$\dot{x}_2 = -3x_1^2(x_1^3 - x_2) + v$$

The derivatives of the transformed coordinates are calculated as

$$\dot{z}_1 = -x_1^2 + x_2 = z_2$$
$$\dot{z}_2 = -3x_1^2 \cdot \dot{x}_1 + \dot{x}_2 = v$$

Therefore we can conclude that the system is linear and controllable in the transformed coordinates with the calculated feedback.

10.3 Input–output Linearization

As we have seen in the previous section, the conditions of the solvability of the State-space Exact Linearization Problem in Theorem 10.2.1 are strict and, in general, the PDE (10.44) is difficult to solve analytically. Therefore the exact linearization may not be applicable or may not be feasible in practical cases. Input–output linearization is an alternative way of achieving linear behavior of a system by nonlinear coordinate transformation.

Recall from Subsection 5.5.1 that any input-affine state-space system model with relative degree r (with $r \le n$) can be transformed into normal form described in Equations (5.53) where the first r transformed coordinates form a linear and controllable subsystem. One can then calculate u such that $\dot{z}_r = v$. This gives

$$u = \frac{1}{a(z)}(-b(z) + v) \tag{10.67}$$

and

$$
\begin{aligned}
\dot{z}_1 &= z_2 \\
\dot{z}_2 &= z_3 \\
&\cdots \\
\dot{z}_{r-1} &= z_r \\
\dot{z}_r &= v \\
\dot{z}_{r+1} &= q_{r+1}(z) \\
&\cdots \\
\dot{z}_n &= q_n(z) \\
y &= z_1
\end{aligned}
\tag{10.68}
$$

This feedback decomposes the system into two parts:

- a linear subsystem of order r which is influenced by the chosen input u, and
- a nonlinear subsystem described by the zero dynamics.

This observation is stated in the theorem below.

Theorem 10.3.1 (Input–output linearization). *Consider a nonlinear system having relative degree r at x^0. The state feedback*

$$u = \frac{1}{L_g L_f^{r-1} h(x)}(-L_f^r h(x) + v) \qquad (10.69)$$

transforms this system into a system whose input–output behavior is identical to that of a linear system having a transfer function

$$H(s) = \frac{1}{s^r}$$

It is important to observe that linearizing feedback of the input–output linearization determined by Equation (10.69) is easy to compute and influences only the r coordinates of the system. The rest, which is not influenced by the feedback, is determined by the zero dynamics of the system (see in Subsection 5.5.1). Therefore, *the main applicability condition of input–output linearization is to have a stable zero dynamics* in a wide domain of the state-space, or even better, a globally stable zero dynamics.

Example 10.3.1 (Input–output linearization)

Consider again the system (10.66), but now with the predefined output equation $h(x) = x_2$. It's easy to check that $L_g h(x) = 1$ in each point of the state-space and thus the system has uniformly relative degree 1 with respect to this output. The components of the feedback for the input–output linearization are

$$\alpha(x) = \frac{-L_f h(x)}{L_g h(x)} = x_1 x_2, \quad \beta(x) = 1$$

The state equations of the closed-loop system are

$$\dot{x}_1 = -x_1^3 + x_2$$
$$\dot{x}_2 = -x_1 x_2 + x_1 x_2 + v = v$$

10.4 Process Systems with Maximal Relative Degree

The following simple process example illustrates the technique of feedback linearization and introduces the general statements on process systems with maximal relative degree which follow thereafter.

Example 10.4.1 (Feedback linearization of a tank)
Tank with outflow valve example

Let us have a simple tank with output convection driven by the hydrostatic pressure at the bottom of the tank through a valve with square root characteristics. Assume that the inflow to the tank is the input variable. Develop the nonlinear state-space model of the tank and apply feedback linearization.

The state equation of the tank is the overall mass balance over the tank–valve system:

$$\rho A \frac{dh}{dt} = \rho v_B - k^* \sqrt{\rho g h} \tag{10.70}$$

where ρ is the density, A is the uniform cross-section, h is the level, v_B is the volumetric inflow, g is the gravitation constant and k^* is the valve constant.

The above equation can be easily rewritten in the usual state equation form by assuming $x = h$ and $u = v_B$:

$$\frac{dx}{dt} = -k_2 \sqrt{x} + k_1 u \tag{10.71}$$

The solution of the State-space Exact Linearization Problem is the following:

- $g(x) = k_1$ is non-zero for any $x \in \mathbb{R}$ because it is a constant function. The distribution span$\{g\}$ is involutive since it is one-dimensional.
- In this case we have to deal only with the non-triviality condition
  $L_g \lambda(x^0) \neq 0$. It is satisfied at any point of the state-space if we choose λ, *e.g.* as follows

  $$\lambda(x) = x \tag{10.72}$$

  since $L_g \lambda(x) = k_1$. Note that this means that the State-space Exact Linearization Problem can be solved in any point of the state-space.
- Thus

  $$\alpha(x) = \frac{-L_f \lambda(x)}{L_g \lambda(x)} = \frac{k_2 \sqrt{x}}{k_1} \tag{10.73}$$

  and

  $$\beta(x) = \frac{1}{L_g \lambda(x)} = \frac{1}{k_1} \tag{10.74}$$

- The linearizing transformation is given by

$$z = \Phi(x) = \text{col}(\lambda(x)) = x \tag{10.75}$$

The state equation of the system after applying the linearizing feedback is written as

$$\dot{x} = -k_2\sqrt{x} + k_1\left(\frac{k_2\sqrt{x}}{k_1} + \frac{1}{k_1}v\right) = v \tag{10.76}$$

which is linear and controllable.

There are several characteristic process systems which exhibit the maximal relative degree property $r = n$. In order to find them we use the notion of the *structure graph* of a lumped system.

Definition 10.4.1 (Structure graph)
The structure graph of a lumped nonlinear system is a directed graph $G = (V, E)$ where

- *the vertex set V is partitioned into three subsets*

$$V = U \cup X \cup Y \quad U \cap X = X \cap Y = U \cap Y = \emptyset$$

  with U being the set of inputs, X the set of state and Y the set of output variables,
- *the edge set E describes direct influences between the variables, i.e. a directed edge (v_i, v_j) exists from v_i to v_j if the variable v_i is present on the right-hand side of the equation determining v_j (in the case of output) or $\dot{v}_j$ (in the case of state).*

Note that there is no inward directed edges for input and no outward directed edges for output variables in the graph.

Note that there is a relationship between structure matrices introduced in Subsection 7.4.1 and the structure graph above (see [33]).

A necessary (but not sufficient) structural condition for a SISO system to have relative degree $r = n$ is that the shortest path (or paths) from the input to the output should traverse all the state variables.

The following cases with clear engineering meaning satisfy the condition above.

1. *Systems with only a single state variable ($n = 1$)*
   The conditions are trivially fulfilled but it is not a practically interesting case (see Example 10.4.1).

2. *Cascade systems (lumped DPS systems or a sequence of CSTRs) with a single balance in each lump and with special input and output*
   An example of this is a sequence of isothermal CSTRs with constant overall mass and a simple r-th order reaction in each tank where the component mass balances form the set of state equations:

$$\frac{dc_A^{(i)}}{dt} = \frac{v}{V^{(i)}}(c_A^{(i-1)} - c_A^{(i)}) - kc_A^{(i)r}$$

If we then choose $u = c_A^{(0)}$ as the inlet concentration and $y = c_A^{(n)}$ as the outlet concentration, then the structure of the input and output functions is given by:

$$g(x) = \begin{bmatrix} * \\ 0 \\ \cdots \\ 0 \end{bmatrix}, \quad h^T(x) = \begin{bmatrix} 0 \\ \cdots \\ 0 \\ ** \end{bmatrix}$$

where $*$ denotes a non-zero entry, and the system is of full relative degree.

3. *Cascade systems with more than one balance but with a special selection of input and output variables*
   An example is a heat exchanger cell where the cold side flow rate is manipulated ($u = v_c$) and the hot side outlet temperature is chosen as the output variable $y = T_{ho}$.

10.5 Exact and Input–output Linearization of a Continuous Fermenter

In this section we illustrate the use of feedback and input–output linearization in the example of a simple continuous fermenter described in Section 4.5.3 following [76].

The nonlinear state-space model of the fermenter is given in Equations (4.78)–(4.79). The variables and parameters of the fermentation process model are collected in Table 4.1.

A nonlinear technique, feedback linearization is applied to the nonlinear state-space model for changing the system dynamics into a linear one. Then, different linear controllers are employed in the linearized system.

10.5.1 Exact Linearization via State Feedback

In order to satisfy the conditions of exact linearization, first we have to find an artificial output function $\lambda(x)$ that is a solution of the PDE:

$$L_g\lambda(x) = 0 \tag{10.77}$$

i.e.

$$\frac{\partial \lambda}{\partial x_1} g_1(x) + \frac{\partial \lambda}{\partial x_2} g_2(x) = 0 \tag{10.78}$$

It's easy to check that

$$\Phi\left(\frac{V(-S_F + x_2 + S_0)}{x_1 + X_0}\right) \tag{10.79}$$

is a solution of (10.78) where Φ is an arbitrary continuously differentiable (C^1) function. Let us choose the simplest possible output function, *i.e.*

$$\lambda(x) = \frac{V(-S_F + x_2 + S_0)}{x_1 + X_0} \tag{10.80}$$

Then the components of state feedback $u = \alpha(x) + \beta(x)v$ for linearizing the system are calculated as

$$\alpha(x) = \frac{-L_f^2 \lambda(x)}{L_g L_f \lambda(x)} \tag{10.81}$$

$$\beta(x) = \frac{1}{L_g L_f \lambda(x)} \tag{10.82}$$

and the new coordinates are

$$z_1 = \lambda(x) = \frac{V(-S_F + x_2 + S_0)}{x_1 + X_0}$$

$$z_2 = L_f \lambda(x)$$

$$= \frac{\mu_{max} V(S_0 X_0 + S_0 x_1 + X_0 x_2 + x_1 x_2 - Y S_F S_0 - Y S_F x_2 + Y S_0^2 + 2Y x_2 S_0 + Y x_2^2)}{Y(x_1 + X_0)(K_2 x_2^2 + 2K_2 x_2 S_0 + K_2 S_0^2 + x_2 + S_0 + K_1)}$$

The state-space model of the system in the new coordinates is

$$\dot{z}_1 = z_2 \tag{10.83}$$
$$\dot{z}_2 = v \tag{10.84}$$

which is linear and controllable.

The exactly linearized model may seem simple but if we have a look at the new coordinates we can see that they are quite complicated functions of x depending on both state variables. Moreover, the second new coordinate z_2 depends on μ, which indicates that the coordinate transformation is sensitive with respect to uncertainties in the reaction rate expression.

10.5.2 Input–output Linearization

Here we are looking for more simple and practically useful forms of linearizing the input–output behavior of the system. The static nonlinear full state feedback for achieving this goal is calculated as

$$u = \alpha(x) + \beta(x)v = -\frac{L_f h(x)}{L_g h(x)} + \frac{1}{L_g h(x)} v \tag{10.85}$$

provided that the system has relative degree 1 in the neighborhood of the operating point where v denotes the new reference input. As we will see, the key point in designing such controllers is the selection of the output (h) function where the original nonlinear state equation (4.78) is extended by a nonlinear output equation $y = h(x)$ where y is the output variable.

Controlling the Biomass Concentration. In this case

$$h(x) = \bar{X} = x_1$$

and

$$\alpha(x) = -\frac{L_f h(x)}{L_g h(x)} = \frac{V\mu_{max}(x_2 + S_0)}{K_2(x_2 + S_0)^2 + x_2 + S_0 + K_1} - F_0 \tag{10.86}$$

$$\beta(x) = -\frac{V}{x_1 + X_0} \tag{10.87}$$

The outer loop for stabilizing the system is the following:

$$v = -k \cdot h(x) \tag{10.88}$$

It was found that the stabilizing region of this controller is quite wide but not global. The time derivative of the Lyapunov function is shown in Figure 10.1.

Controlling the Substrate Concentration. In this case, the chosen output is

$$h(x) = \bar{S} = x_2$$

The full state feedback is composed of the functions

$$\alpha(x) = -\frac{V\mu_{max}(x_2 + S_0)(x_1 + X_0)}{Y(K_2(x_2 + S_0)^2 + x_2 + S_0 + K_1)(S_F - x_2 - S_0)} + F_0 \tag{10.89}$$

$$\beta(x) = \frac{V}{S_f - x_2 - S_0} \tag{10.90}$$

In the outer loop, a negative feedback with gain $k = 0.5$ was applied, *i.e.* $v = -0.5x_2$. The time derivative of the Lyapunov function of the closed-loop system as a function of $\bar{X}$ and $\bar{S}$ is shown in Figure 10.2.

Note that for this case it was proven (see the zero dynamics analysis of the continuous fermenter in Subsection 5.5.2) that the closed-loop system is globally stable except for the singular points, where the biomass concentration is zero.

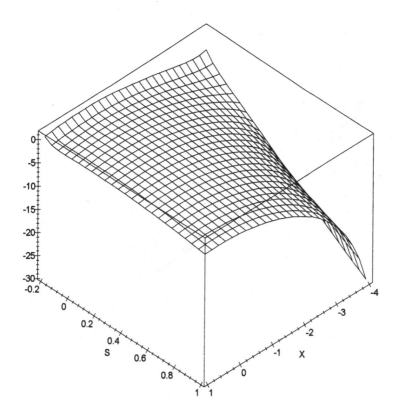

Figure 10.1. Time derivative of the Lyapunov function as a function of centered state variables $q_1 = 1, q_2 = 1$, linearizing the biomass concentration, $k = 0.5$

Controlling the Linear Combination of the Biomass and the Substrate Concentrations. In this case, the output of the system was chosen as

$$h(x) = Kx \tag{10.91}$$

where the row vector K is calculated as the result of the previously described LQR cheap control design problem in Subsection 9.5.2. Then the functions α and β are given as

$$\alpha(x) = -\frac{Kf(x)}{Kg(x)} \tag{10.92}$$

$$\beta(x) = \frac{1}{Kg(x)} \tag{10.93}$$

The value of K was $[-0.6549\ \ 0.5899]^T$ and in the outer loop a negative feedback with gain $k = 0.5$ was applied. The time derivative of the Lyapunov function of the closed-loop system is shown in Figure 10.3.

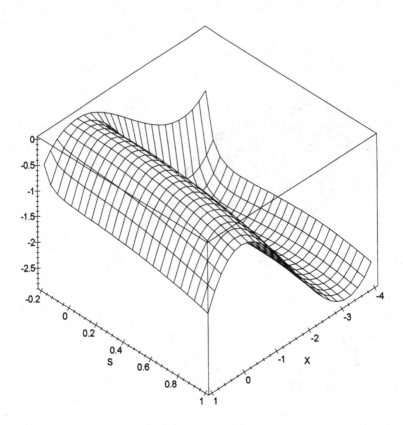

Figure 10.2. Time derivative of the Lyapunov function as a function of centered state variables $q_1 = 1, q_2 = 1$, linearizing the substrate concentration, $k = 0.5$

10.6 Output Selection for Feedback Linearization

The problem of output selection for nonlinear output feedback is a general issue for all nonlinear systems (see, *e.g.* [81]). In several cases, an unstable nonlinear system can be stabilized by appropriate controls provided that a proper output is selected.

In Section 10.2 it was shown that exact linearization has two main disadvantages: first, partial differential equations have to be solved to calculate the output function h, and second, if we manage to find an appropriate output then it is (together with the coordinates transformation Φ) often a highly nonlinear function of the state variables, which is hard to treat practically. Therefore, it is of interest to select a *linear* output with advantageous properties from an engineering point of view.

In Section 5.5.2 we managed to show that controlling the substrate concentration is enough for the global stabilization of continuous fermentation processes (*i.e.* we were able to select an output with globally asymptotically stable zero dynamics). However, the analytical investigation of the stability

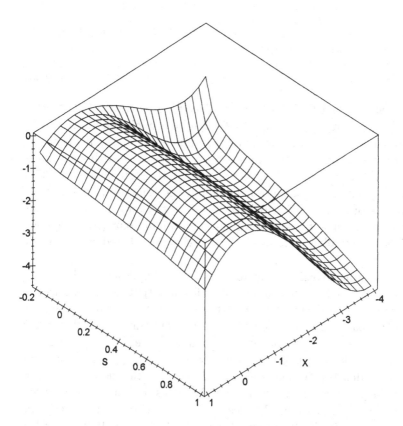

Figure 10.3. Time derivative of the Lyapunov function as a function of centered state variables $q_1 = 1, q_2 = 1$, linearizing the linear combination of the biomass and substrate concentrations, $K = [-0.6549 \ \ 0.5899]^T$, $k = 0.5$

of the zero dynamics is usually not possible in the higher dimensional cases. Nevertheless, using the well-known results of optimal control and the theory of nonlinear systems, a generally applicable linear output selection method can be derived.

To establish a relation between linear systems and the zero dynamics of nonlinear systems, let us consider a single-input linear system given in standard state-space form:

$$\dot{x} = Ax + Bu \tag{10.94}$$

where $x \in \mathbb{R}^n$, $u \in \mathbb{R}$, A and B are real matrices of appropriate dimensions. Let us assume that all state variables are measurable. Furthermore, assume that the pair (A, B) is controllable.

Consider a full state feedback $u = -Kx$ (possibly resulting from a linear controller design method, *e.g.* pole-placement or LQ-design) that asymptotically stabilizes the system (10.94).

The controlled closed-loop system can be interpreted in such a way that the system (10.94) with the output equation

$$y = Kx \tag{10.95}$$

is fed back with output feedback $u = -y$. Therefore the system (10.94) with the special linear output equation (10.95) can be a minimum-phase system (*i.e.* the real parts of the zeros of its transfer function are strictly negative). In the linear case, the zero dynamics is a linear dynamics with eigenvalues coinciding with the zeros of the transfer function of the system (see, *e.g.* [37]). It is also true that the linear approximation at the equilibrium point of the zero dynamics of a system (3.19) is the same as the zero dynamics of the linear approximation of the system at the equilibrium point (see Chapter 4 in [37]). These facts allow us to search for linear outputs based on the locally linearized system.

The resulting feedback gain vector K_{LQ} of an LQ-design problem (see Section 9.3) is generally a good candidate for linear output selection, since the advantageous phase and gain margins of the LQ-loop are well-known (see, *e.g.* [22] or [48]). In particular, the infinite gain margin guarantees that the system (10.94) with output K_{LQ} will be locally minimum-phase. The explanation of this fact is the following: it is well-known from classical root-locus analysis that as the feedback gain increases towards infinity, the closed-loop poles migrate to the positions of open-loop zeros [64]. Thus the infinite-gain margin in the linear case means that the closed-loop system remains stable if we increase the loop-gain. It also means that the linear system (10.94) together with the output equation

$$y = K_{LQ} \cdot x \tag{10.96}$$

has stable zeros (*i.e.* asymptotically stable zero dynamics).

By applying the above-mentioned relation between the zero dynamics of the locally linearized system and the locally linearized zero dynamics of the original nonlinear system, we obtain that the original nonlinear model with the same linear output (10.96) will have at least locally asymptotically stable zero dynamics. Note that this property is not necessarily true in the case of the output generated by any locally stabilizing linear controller, but it is always fulfilled by the feedback gain of LQ-controllers. In summary, the method has two main advantages:

- it can be applied directly to high-dimensional nonlinear systems without complicated calculations,
- the closed-loop performance can be evaluated and set easily because the output is linear.

Output Selection for the Continuous Fermenter. Now, let us check the minimum-phase property of the linearized system for continuous fermentation processes. Consider the linearized model in Equations (4.84)–(4.87) together with the output equation in Equation (10.96), where we examine two feedback gains in Subsection 9.5.2, namely:

1. $K_{LQ_1} = [-0.6549 \ 0.5899]$
2. $K_{LQ_2} = [-1.5635 \ 2.5571]$

where the LQ-controller design parameters were $Q = I^{2\times2}$, $R = 1$ and $Q = 10 \cdot I^{2\times2}$, $R = 1$ respectively.

The zeros of the corresponding transfer functions are:

1. $z_1 = -0.6590$ and
2. $z_2 = -0.7083$

which are stable. Thus both of the outputs generated by Equation (10.96) can be applied to feedback linearization of the continuous fermenter.

10.7 Further Reading

The use of the normal form (5.53) was first proposed in [36]. The state-space exact linearization problem was proposed and solved in [15] for single-input systems and [38] for multi-input systems. The notion of zero dynamics is from [17] and it was used for stabilization in [18].

10.8 Summary

Feedback linearization techniques, namely exact linearization and input–output linearization have been introduced in this chapter together with the related basic notion of relative degree. The existence of the solution and the applicability conditions of both the exact and input–output linearization have also been described and analyzed.

Engineering conditions of process systems that are of maximal relative degree have also been discussed. The linearization techniques have been illustrated in a case study of a simple continuous fermenter. Finally, a practical way of output selection for feedback linearization has also been proposed.

10.9 Questions and Application Exercises

Exercise 10.9.1. Compare the exact linearization and input–output linearization techniques for input-affine nonlinear systems. Comment on their applicability conditions and their need for computational resources.

Exercise 10.9.2. Give engineering conditions for a process system to have maximal relative degree.

Exercise 10.9.3. Compute the relative degree of the input-affine nonlinear state-space model of the simple fed-batch fermenter developed in Subsection 4.5.2 and given in Equations (4.71)–(4.74) with the output equation and input variable:

$$y = x_1 = X, \quad u = S_f$$

Exercise 10.9.4. Consider the bilinear (nonlinear) input-affine state-space model of the heat exchanger cell developed in Section 4.4.4 with the input and output variables

$$u = v_c, \quad y = T_h = x_2$$

Check the applicability conditions and then perform exact linearization of the system.

Exercise 10.9.5. Consider the bilinear (nonlinear) input-affine state-space model of the heat exchanger cell developed in Section 4.4.4 with the input and output variables

$$u = v_c, \quad y = T_c = x_1$$

Check the applicability conditions and then perform input–output linearization of the system.

Exercise 10.9.6. Consider a simple CSTR model developed in Subsection 4.5.1 with the input and output variables

$$u = \bar{c}_{Ain}, \quad y = \bar{c}_A$$

Compute the relative degree of the system and apply exact linearization if it is possible.

Exercise 10.9.7. Consider the following two-dimensional nonlinear system model (the same as in Exercise 6.9.4) in its input-affine state-space form:

$$\frac{dx_1}{dt} = x_1^{1/2} x_2^{3/2} - x_1 x_2^3 + x_1^{1/2} u_1$$

$$\frac{dx_2}{dt} = 3x_1 x_2^{1/2} - 2x_1^2 x_2$$

$$y = x_2^2$$

Compute the relative degree of the system and apply exact linearization if it is possible.

Exercise 10.9.8. Consider again the above two-dimensional nonlinear system model (the same as in Exercise 6.9.4) but with another output in its input-affine state-space form:

$$\frac{dx_1}{dt} = x_1^{1/2} x_2^{3/2} - x_1 x_2^3 + x_1^{1/2} u_1$$

$$\frac{dx_2}{dt} = 3x_1 x_2^{1/2} - 2x_1^2 x_2$$

$$y = x_1$$

Check the applicability conditions and then design an input–output linearizing feedback to the above system.

11. Passivation by Feedback

A conceptually simple but yet powerful way of obtaining a passive system from a non-passive one is to apply a nonlinear static feedback to make it passive. This idea is the basis of passivation by feedback, which is the subject of this chapter.

The material is presented in the following sections:

- *The passivation problem and static feedback design*
  Here we describe the necessary and sufficient conditions that enable us to construct a static feedback which makes a nonlinear input-affine system passive.
- *Stabilization using control Lyapunov functions*
  A special case of stabilizing static feedback design is when a suitable chosen Lyapunov function is prescribed for the closed-loop system. This practically important controller design method is the subject of this section.
- *Direct passivation of a continuous fermenter*
  This simple case study illustrates the notions and tools of direct passivation by feedback.
- *Direct passivation of a gas turbine*
  This section presents a more detailed case study of practical importance where the controller based on passivation is the key element of control system design.

11.1 The Passivation Problem and Static Feedback Design

In this section, we examine the conditions of transforming an input-affine non-passive system by static nonlinear feedback of the form (9.3), into a system which is passive with respect to the new input and the system's output. The material covered in the section is based on [19] and [80].

Consider the input-affine state-space model (3.19) in the form of

$$\Sigma : \quad \begin{aligned} \dot{x}(t) &= f(x(t)) + \sum_{i=1}^{m} g_i(x(t))u_i(t) \\ y(t) &= h(x(t)) \end{aligned}$$

and let us denote it simply by Σ. With the static state feedback with external input as defined in Definition 9.1.2 in the form

$$u = \alpha(x) + \beta(x)v$$

the closed-loop system denoted by $\Sigma_{\alpha,\beta}$ is given by

$$\Sigma_{\alpha,\beta} : \quad \begin{array}{l} \dot{x} = [f(x) + g(x)\alpha(x)] + g(x)\beta(x)v \\ y = h(x) \end{array} \qquad (11.1)$$

Definition 11.1.1 (Feedback equivalence)
Σ *is said to be feedback-equivalent to $\Sigma_{\alpha,\beta}$ in Equation (11.1).*

Now, we present the main line of the solution without going too much into the details. First, suppose that we have achieved our goal, *i.e.* that $\Sigma_{\alpha,\beta}$ is passive with a storage function $S \geq 0$. Then

$$S(x(t_1)) - S(x(t_0)) \leq \int_{t_0}^{t_1} v^T(t)y(t)dt, \quad \forall t_1 > t_0 \qquad (11.2)$$

Now, examine the behavior of the closed-loop system when its output is identically zero:

$$\Sigma_{\alpha,\beta}^c : \quad \begin{array}{l} \dot{x} = [f(x) + g(x)\alpha(x)] + g(x)\beta(x)v \\ 0 = h(x) \end{array} \qquad (11.3)$$

Then it is easy to see from (11.2) that

$$S(x(t_1)) - S(x(t_0)) \leq 0, \quad \forall t_1 > t_0 \qquad (11.4)$$

which means that *the zero dynamics of the closed-loop system is stable.*

Let us now return to the original open-loop system system with the constraint $y = 0$

$$\Sigma^c : \quad \begin{array}{l} \dot{x} = f(x) + g(x)u \\ 0 = h(x) \end{array} \qquad (11.5)$$

and assume that $(x(t), u(t))$ is a solution of Σ^c. Then, by expressing v from the feedback law, it is clear that

$$x(t), \quad v(t) = \beta^{-1}(x(t))(u(t) - \alpha(x(t))) \qquad (11.6)$$

is a solution of $\Sigma_{\alpha,\beta}^c$, and (11.4) holds in this case, too. This way, we have concluded that *the stability of the zero dynamics is a necessary condition of passivation.*

Now, let us calculate the input which is necessary to keep the system's output at 0. If we consider the condition $y = 0$, then naturally $\dot{y} = 0$ and we can write

$$\dot{y} = h_x(x)f(x) + h_x(x)g(x)u \qquad (11.7)$$

This gives the condition below for u as the stabilizing (or passivating) input.

Definition 11.1.2 (Passivating feedback)
The static state feedback of the form

$$u^*(x) = -[h_x(x)g(x)]^{-1}h_x(x)f(x) \tag{11.8}$$

is called the passivating feedback. Note that $h_x(x)g(x)$ must be of full rank in a neighborhood of 0 in order to calculate the inverse in (11.8).

Of course, the initial condition of the system must be in the set
$\mathcal{X}_c = \{x \in \mathcal{X} | h(x) = 0\}$.

To summarize the above, let us state the following theorem:

Theorem 11.1.1 (Necessary condition of passivation). *Suppose Σ is feedback-equivalent to a passive system with a C^2 storage function S locally about $x = 0$, for which $S(0) = 0$ and $S(x) > 0$, $x \neq 0$. Assume that $h_x(x)g(x)$ has rank m in a neighborhood of 0. Then*

$$S_x(x)[f(x) + g(x)u^*(x)] \leq 0, \quad x \in \mathcal{X}_c \tag{11.9}$$

with $u^*(x)$ defined by (11.8).

The following sufficient condition is very similar to the previous one.

Theorem 11.1.2 (Sufficient condition for passivation). *Consider the system Σ. Suppose $\mathrm{rank}(h_x(0)g(0)) = m$, and let $S \geq 0$ be such that $S(0) = 0$, $S(x) > 0$, $x \neq 0$ and satisfies $S_x(x)[f(x) + g(x)u^*(x)] \leq 0$, $x \in \mathcal{X}_c$. Then Σ is locally feedback-equivalent to a passive system with a storage function which is positive definite at $x = 0$.*

The complete proof can be found in [80]. The idea of the proof is to describe the system as a standard feedback interconnection of two passive systems using local coordinates transformations.

11.2 Stabilization Using Control Lyapunov Functions

In this section we investigate the possibilities of stabilizing a nonlinear system using a pre-defined Lyapunov function. Consider an input-affine nonlinear system (3.19) and the set of static feedbacks (without new input) introduced in Definition 9.1.1. Assume that the system is stabilized by a smooth feedback $u = \alpha(x)$. Then, according to the converse Lyapunov theorem (see in Theorem 7.3.2), there exists a proper Lyapunov function $V : \mathcal{X} \mapsto \mathbb{R}^+$, which is decreasing along the closed-loop system's trajectories, i.e.

$$\frac{d}{dt}V = \frac{\partial V}{\partial x}(f(x) + g(x)\alpha(x)) = L_f V(x) + \alpha(x)L_g V(x) < 0 \tag{11.10}$$

It is visible from (11.10) that if $L_g V(x)$ is identically set to zero, then necessarily $L_f V(x) < 0$.

Motivated by this, let us define a special function that can be the Lyapunov function of the closed-loop system.

Definition 11.2.1 (Control Lyapunov function)

A control Lyapunov function of the system (3.19) is a proper function
$V : \mathcal{X} \mapsto \mathbb{R}^+$ *with the property*

$$L_g V(x) = 0 \quad \Rightarrow \quad L_f V(x) < 0, \quad \forall x \neq 0 \tag{11.11}$$

The importance of control Lyapunov functions is that the existence of such
a function is also a sufficient condition for a stabilizing feedback law, as is
stated by the following theorem:

Theorem 11.2.1 (Artstein-Sontag's theorem). *Consider a nonlinear input-affine system of the form (3.19) where f and g are smooth vector fields and $f(0) = 0$. There exists a feedback law $u = \alpha(x)$ which is smooth on $\mathbb{R}^n \backslash \{0\}$, continuous at $x = 0$ and globally asymptotically stabilizes the equilibrium $x = 0$ if and only if there exists a function $V : \mathcal{X} \mapsto \mathbb{R}^+$ with the following properties:*

1. *V is a control Lyapunov function for the system (3.19).*
2. *When $\forall \epsilon > 0$ there exists $\delta > 0$ such that for $\|x\| < \delta$, $x \neq 0$ there is some u, $|u| < \epsilon$ such that*

$$L_f V(x) + L_g V(x)u < 0$$

Property 2 is required for the continuity of the feedback law at 0.

The proof of this theorem is constructive (it can be found, *e.g.* in [37]) and
gives a possible stabilizing feedback law based on control Lyapunov functions
in the following form:

$$\alpha(x) = \begin{cases} 0 & \text{if } L_g V(x) = 0 \\ -\dfrac{L_f V(x) + \sqrt{(L_f V(x))^2 + (L_g V(x))^4}}{L_g V(x)} & \text{otherwise} \end{cases} \tag{11.12}$$

Quadratic Lyapunov functions are popular and powerful Lyapunov function candidates when analyzing stability of nonlinear systems. The notion of
quadratically stabilizable systems given by Definition 7.3.2 reflects this fact
such that a special class of nonlinear systems is formed which enables us to
have a quadratic Lyapunov function.

The following example illustrates how to construct a quadratic Lyapunov
function for a simple nonlinear system:

Example 11.2.1 (Quadratic control Lyapunov function)

The aim in this example is to find a control Lyapunov function for the following input-affine system:

$$
\begin{bmatrix} \dot{x}_1 \\ \dot{x}_2 \\ \dot{x}_3 \end{bmatrix} = \begin{bmatrix} 2x_1 + x_2^2 + x_3 \\ x_1 x_2 - 2x_2 \\ x_1^2 x_3 \end{bmatrix} + \begin{bmatrix} 1 \\ 0 \\ 0 \end{bmatrix} u \tag{11.13}
$$

We seek the control Lyapunov function V in the following simple form:

$$
V(x) = \frac{1}{2}(q_1 x_1^2 + q_2 x_2^2 + q_3 x_3^2)
$$

where q_1, q_2 and q_3 are positive real parameters.
The gradient of V is then

$$
V_x(x) = \begin{bmatrix} q_1 x_1 & q_2 x_2 & q_3 x_3 \end{bmatrix}
$$

The condition $L_g V(x) = 0$ now specializes to $q_1 x_1 = 0$, which is equivalent to $x_1 = 0$.
Now, we calculate the sign of the derivative of the candidate Lyapunov function along the system trajectories restricted to $x_1 = 0$.

$$
L_f V(x) = \begin{bmatrix} q_1 x_1 & q_2 x_2 & q_3 x_3 \end{bmatrix} \begin{bmatrix} 2x_1 + x_2^2 + x_3 \\ x_1 x_2 - 2x_2 \\ x_1^2 x_3 \end{bmatrix}
$$

$$
= q_1 x_1(2x_1 + x_2^2 + x_3) + q_2 x_2(x_1 x_2 - 2x_2) + q_3 x_3(x_1^2 x_3) = -2q_2 x_2^2
$$

It is clear from the above that V is a control Lyapunov function of the system for any $q_1, q_2, q_3 > 0$.

11.3 Control Lyapunov Function of a Continuous Fermenter

In this section we investigate a simple quadratic Lyapunov function, *i.e.* whether it can be the control Lyapunov function of the continuous fermentation process described in Subsection 4.5.3. The results of the forthcoming calculations are based on the parameter values listed in Subsection 4.5.2 and the operating point is the optimal one calculated in Subsection 4.5.3. We will use the centered version of the state equations with f and g in (4.79) with the following notations:

$$x_1 = \bar{X}, \; x_2 = \bar{S}, \; x_1^* = X_0, \; x_2^* = S_0, \; u^* = F_0$$

Let the control Lyapunov function candidate be given in the following simple form:

$$V_c(x) = \frac{1}{2}(q_1 x_1^2 + q_2 x_2^2), \quad q_1, q_2 > 0 \tag{11.14}$$

It's easy to see that the set $L_g V_{c0} := \{x \in \mathbb{R}^2 \mid L_g V_c(x) = 0\}$ is an ellipse in the state-space, since

$$L_g V_c(x) = -\frac{1}{V}(q_1 x_1^2 + q_1 x_1 x_1^* + q_2 x_2(x_2^* - S_F) + q_2 x_2^2) \tag{11.15}$$

Let us choose $q_1 = 1$ and $q_2 = 1$ in (11.14). Then (11.15) defines a circle in the state-space as is shown in Figure 11.1. To check the sign of the derivative of

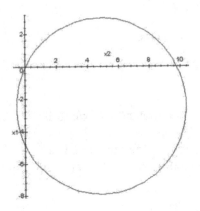

Figure 11.1. The set $\{x \in \mathbb{R}^2 \mid L_g V_c(x) = 0\}$ for $q_1 = 1$ and $q_2 = 1$.

the control Lyapunov function candidate, we first express the x_1 coordinate as a function of x_2 from $L_g V_c(x) = 0$, *i.e.* we try to solve this equation for x_1. The solutions are

$$x_1^{(1,2)} = \frac{-x_1^* \pm \sqrt{(x_1^*)^2 - 4x_2 x_2^* + 4x_2 S_F - 4x_2^2}}{2} \tag{11.16}$$

which define the upper and lower halves of the circle in Figure 11.1. Let us first check the derivative of the control Lyapunov function candidate along f. For this, first calculate $L_f V$, which reads

$$L_f V_c(x) = x_1 f_1(x) + x_2 f_2(x) = x_1 \left(\mu(x_2 + x_2^*)(x_1 + x_1^*) - \frac{(x_1 + x_1^*)u^*}{V} \right)$$

$$+ x_2 \left(-\frac{\mu(x_2 + x_2^*)(x_1 + x_1^*)}{Y} + \frac{(S_F - x_2 - x_2^*)u^*}{V} \right) \tag{11.17}$$

Now check the sign of $L_f V_c$ first in the upper half-circle by substituting $x_1^{(1)}$ into (11.17); we then get a function with only one independent variable (x_2). Instead of algebraic derivation (in fact, this will be the reader's exercise at the end of the chapter), we only give the graph of this function in Figure 11.2. It is visible that the sign of the Lie-derivative is negative in a neighborhood of x_2 for $x_2 \neq 0$. (It can be shown that the function is negative on the whole half-circle for $x_2 \neq 0$.)

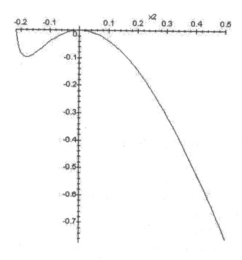

Figure 11.2. The derivative of the control Lyapunov candidate along f restricted to the upper half-circle

Next let us examine the sign of $L_f V_c$ in the lower half-circle. For this, we substitute x_1^2 into (11.17). Similarly to the previous case, we only present the graph of the obtained function to illustrate the solution in Figure 11.3. As is visible for $x_2 > 0$, the sign of the function is positive, but this region would correspond to an operating region where $X < 0$ (see also Figure 11.1 again), which is practically not meaningful. Furthermore, it can be shown that the sign of $L_f V_c$ is again negative in the lower half-circle if $X > 0$.

Finally, we can conclude that the proposed simple quadratic candidate V_c with $q_1 = q_2 = 1$ is a control Lyapunov function of the continuous bio-reactor model in the physically meaningful operating region. Now, it is possible to apply Theorem 11.2.1 to design a stabilizing feedback of the form (11.2).

11.4 Case Study: Direct Passivation of a Gas Turbine

As a case study, a simple nonlinear controller for a low-power gas turbine based on direct passivation is introduced here [3]. It uses a nonlinear state-

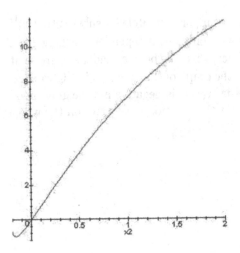

Figure 11.3. The derivative of the control Lyapunov candidate along f restricted to the lower-half circle

space model of the gas turbine in input-affine form based on first engineering principles, which is described in Section 4.6.

The proposed nonlinear controller is based on a prescribed quadratic control Lyapunov function and is able to solve the protection of the gas turbine. The robustness of the closed-loop system with respect to the time-varying parameters is also investigated.

11.4.1 Nonlinear State-space Model

The nonlinear state-space model in its input-affine form can be developed from the model equations based on first engineering principles. For the gas turbine we use the intensive form of the model equations (4.93), (4.98), (4.99), (4.100) and (4.101) described in Section 4.6.5 with the constitutive equations in Section 4.6.6.

System Variables. The state variables of the gas turbine model are then as follows:

$$x = [\, m_{Comb} \;\; p_3 \;\; n \,]^T \tag{11.18}$$

The value of the only *input variable* ν_{fuel} is also constrained by:

$$0.003669480223 \leq \nu_{fuel} \leq 0.02701065149 \; [\text{kg/sec}]$$

The set of possible *disturbances* includes:

$$d = [\, p_1 \;\; T_1 \;\; M_{load} \,]^T \tag{11.19}$$

It is important to observe that the state and disturbance variables can only vary within a prescribed operating region described in Section 4.6.7.

Finally we construct the *set of measurable output variables*:

$$y = [\, T_4 \ \ p_3 \ \ n \,]^T \tag{11.20}$$

Model Equations. The dynamic equations with these vectors can be transformed into input-affine form:

$$\frac{dx}{dt} = f(x) + g(x)u \tag{11.21}$$

$$y = h(x) \tag{11.22}$$

with the following f, g and h functions:

$$f(x) = \begin{bmatrix} f_1(x_1, x_2, x_3, d_1, d_2) \\ f_2(x_1, x_2, x_3, d_1, d_2) \\ f_3(x_1, x_2, x_3, d_1, d_2, d_3) \end{bmatrix} \tag{11.23}$$

$$g(x) = \begin{bmatrix} const \\ const \\ 0 \end{bmatrix} \tag{11.24}$$

$$h(x) = \begin{bmatrix} h_1(x_1, x_2, x_3, d_1) \\ x_2 \\ x_3 \end{bmatrix} \tag{11.25}$$

It is important to note that these functions do not depend on only the state variables, but also on the disturbance vector. Further, observe that $g(x)$ does not depend on the state vector x. This means that the effect of the input is linear to the time derivative of the state vector.

Open-loop Properties of the System. It is found by standard nonlinear analysis ([37]) that the developed model is reachable and observable in the whole application domain and stable in a small neighborhood of any admissible operating point ([2]). Therefore the asymptotic stability of all operating points can be guaranteed in the whole application domain.

The measurable time-varying parameters of the system are collected into the disturbance vector. All elements of this vector are correctly measurable and significantly change their values during the operation of the gas turbine.

There are also some unmeasurable time-varying parameters in the nonlinear model. The value of the parameters σ_I, σ_{Comb}, σ_N and η_{Comb} are not constant, but they change their values only about ± 2 percent.

Because of these properties, we have to achieve disturbance-rejection and robustness of the closed-loop system as stated in the control aims before.

11.4.2 Controller Design

Control Aims. The following control aims is set for our low-power gas turbine:

- The number of revolutions has to follow the position of the throttle and should not be affected by the load and the ambient circumstances (the disturbance vector).
- The temperatures (basically the total temperature after the turbine) and the number of revolutions has to be limited, their values are constrained from above by their maximum values.

Protection of the Gas Turbine. The aim of the protection of the gas turbine is twofold: to avoid too-high temperatures and too-high numbers of revolutions.

1. The turbine outlet temperature y_1 has a maximum value of 938.15 K. If the setpoint for y_1 is higher than its maximum value, then we can increase the number of revolutions (the temperature will then decrease), but the number of revolutions also has a maximum value: 833.33 sec^{-1}.
2. If the position of the throttle specifies a higher value for the number of revolutions x_3 than its maximum, then we do not allow it; the value will be its maximum.

Static Nonlinear Stabilizing Controller. In the first step, a static nonlinear full state feedback is designed to stabilize the system in the whole operating region. For this, let us assume that the point x_0 is an equilibrium point for the system (*i.e.* $f(x_0) = 0$) and define a positive definite storage function

$$V(x) = (x - x_0)^T M(x - x_0) \tag{11.26}$$

where M is an $n \times n$ positive definite symmetric matrix. If we set the input u as

$$u = v_p + v + w \tag{11.27}$$

where

$$v_p = -\frac{L_f V(x)}{L_g V(x)} \tag{11.28}$$

and

$$v = -k \cdot L_g V(x), \quad k > 0 \tag{11.29}$$

then the closed-loop system will be passive with respect to the supply rate $s = w^T y$ with storage function V where $y = L_g V(x)$ and w is the new reference input ([19]). Note that the feedback law in Equation (11.28) has singular points where $L_g V(x) = 0$, therefore x_0 has to be selected carefully to be outside the real operating region. The new reference input w is calculated from the linear shift between x_0 and the setpoint.

Achieving Robustness with a PI-Controller. In order to remove the steady-state error in the number of revolutions caused by the change of the unmeasurable time-varying parameters, a PI-controller (which is itself again a passive system) is applied in the outer loop, *i.e.*

$$w(t) = k_p(y_{3s} - y_3(t)) + k_i \cdot \int_{t_0}^{t} (y_{3s} - y_3(\tau))d\tau \tag{11.30}$$

where y_{3s} is the constant setpoint for the number of revolutions, $y_3(t)$ is the number of revolutions as a function of time, k_p and k_i are the parameters of the PI-controller.

11.4.3 Simulation Results

For the simulations a typical operating point has been selected:

$$x^* = [\, 0.00580436 \ 197250.6068 \ 700 \,]^T \tag{11.31}$$

$$d^* = [\, 101325 \ 288.15 \ 10 \,]^T \tag{11.32}$$

$$u^* = 0.009252624089 \tag{11.33}$$

Simulation of the Open-loop System. The properties of the open-loop system are well demonstrated by the simulation results. Figure 11.4 shows a case where the number of revolutions has been changed to 400 and the system is not able to reach the setpoint, indicating that the open-loop system is not globally asymptotically stable. Figure 11.5 shows that the open-loop system is sensitive with respect to the change of the disturbances: if we change the M_{load} from 10 to 30 then the engine shuts down.

Simulation of the Closed-loop System. In order to achieve appropriate dynamics of the closed-loop system, we have to tune M to be positive definite and symmetric; to select the positive constant k and the parameters of the PI-controller k_i and k_p. The tuning of these parameters was carried out by a trial and error method and the following values were obtained:

$$M = \begin{bmatrix} 10 & 0 & 0 \\ 0 & 1000 & 0 \\ 0 & 0 & 0.001 \end{bmatrix} \tag{11.34}$$

and $k = 0.0005$, $k_i = 0.7$, $k_p = 1.1$.

Figure 11.6 shows the same case as in Figure 11.4 with our nonlinear controller. It indicates that the controller indeed stabilizes the system globally. Figure 11.7 shows that the closed-loop system is not sensitive to the elements of the disturbance vector. If we increase the load, the number of revolutions decreases, but the nonlinear controller is able to set the original operating point.

Figure 11.8 shows a setpoint change and the robustness of the controller. The position of the throttle has been changed from 700 sec^{-1} to 750 sec^{-1} and simultaneously one time-varying parameter has been raised by 2 percent.

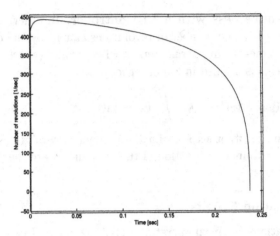

Figure 11.4. Response of the number of revolutions of the open-loop system (unstable)

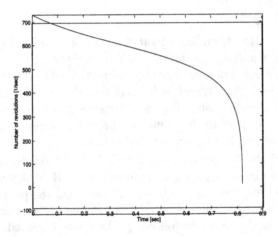

Figure 11.5. Response of the number of revolutions of the open-loop system (sensitive)

Conclusion. The nonlinear controller described above for a low-power gas turbine is able to guarantee the asymptotic stability of the closed-loop system in every operating point of the application domain.

This controller keeps the number of revolutions in accordance with the position of the throttle while the number of revolutions is not affected by the load and ambient circumstances.

At the same, time the controller can protect the gas turbine against a too-high temperature and too-high number of revolutions at a setpoint and the closed-loop system is robust with respect to the time-varying parameters.

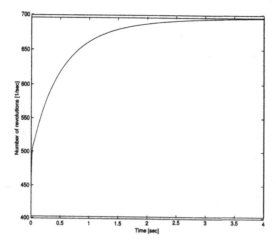

Figure 11.6. Response of the number of revolutions of the closed-loop system (stable)

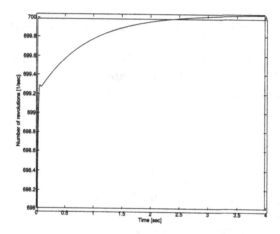

Figure 11.7. Response of the number of revolutions of the closed-loop system (not sensitive)

11.5 Further Reading

The conditions for making a system passive through static nonlinear feedback were derived in [19]. Control Lyapunov functions are described in a rigorous and detailed way in, *e.g.* [68]. The Artstein-Sontag's Theorem for the stabilization of nonlinear system was presented in [67].

Mechatronic systems, such as robots, are the traditional area of passivating control, see, *e.g.* [61] for a recent reference.

Process Systems. The idea of controlling process systems based on thermodynamical principles was introduced in the 1990's [55], [84], [23]. Ydstie

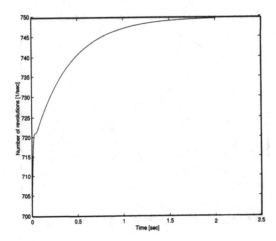

Figure 11.8. Response of the number of revolutions of the closed-loop system (set-point change)

[83] offers a recent survey on passivity-based control of process systems via the second law.

Passivity-based control of nonlinear chemical processes has recently been reported in [63]. The approach has been extended to distributed (*i.e.* infinite dimensional) process systems in [4] and [5], too.

11.6 Summary

The methods for designing static feedback laws that stabilize (or passivate) a nonlinear system are described in this chapter. The practically important method of stabilizing state feedback design uses a prescribed, usually quadratic control Lyapunov function.

Two case studies that illustrate the design and usefulness of stabilizing feedback controllers are presented. The direct passivation of a simple continuous fermenter proposes a controller structure which stabilizes the system globally. The passivation of a gas turbine represents a practically relevant case, where the key element of the controller is the stabilizing loop.

11.7 Questions and Application Exercises

Exercise 11.7.1. Give the design parameters of the passivating and stabilizing controller design. Comment on the effect of these parameters on the properties of the closed-loop system.

Exercise 11.7.2. Give the design parameters of the passivating and stabilizing feedback controller design. Comment on the effect of these parameters on the properties of the closed-loop system.

Exercise 11.7.3. Compare the design and properties of the passivating and stabilizing feedback controllers. Comment on the applicability conditions, the information and model they require and on the computational effort they need.

Exercise 11.7.4. Consider the use of the LTI state-space model of the free mass convection network in Equation (4.22) developed in Subsection 4.2.4. Design a passivating and a stabilizing feedback controller for this simple process system.

Exercise 11.7.5. Consider the LTI state-space model of the heat exchanger cell developed in Subsection 4.4.2 of Section 4.4. Design a passivating and a stabilizing feedback controller for this simple process system.

Exercise 11.7.6. Consider a simple CSTR model developed in Subsection 4.5.1. Construct a passivating feedback controller for to this system.

Exercise 11.7.7. Consider again a simple CSTR model developed in Subsection 4.5.1. Construct a stabilizing feedback controller for this system using a quadratic control Lyapunov function.

Compare the controller with the passivating feedback one designed in Exercise 11.7.6.

Exercise 11.7.8. Consider the bilinear (nonlinear) input-affine state-space model of the heat exchanger cell developed in Subsection 4.4.4 with the input and output variables
$$u := v_c, \quad y = T_h = x_2$$
Construct a stabilizing feedback controller for this system using a quadratic control Lyapunov function.

Exercise 11.7.9. Consider the following two-dimensional nonlinear system model (the same as in Exercise 6.9.4) but with another output in its input-affine state-space form:
$$\frac{dx_1}{dt} = x_1^{1/2} x_2^{3/2} - x_1 x_2^3 + x_1^{1/2} u_1$$
$$\frac{dx_2}{dt} = 3x_1 x_2^{1/2} - 2x_1^2 x_2$$
$$y = x_1$$

Check the applicability conditions and then design a stabilizing feedback controller for the system.

12. Stabilization and Loop-shaping

This chapter deals with stabilizing control and loop-shaping of Hamiltonian systems. The basic notions and tools of the Hamiltonian system models together with process examples were given in Chapter 8.

In the first part of the chapter we follow [21]. Thereafter the methods and results are extended to process systems.

The following sections contain the material of the chapter:

- *Stabilization of Hamiltonian systems*
  The stabilization is done in two logical steps: stable systems are made asymptotically stable by adding damping to the systems while unstable systems are made stable by shaping their potential energy. This leads to a nonlinear proportional-derivative (PD) output feedback controller.
- *Stabilization and loop-shaping of nonlinear process systems*
  Using the specialities of the Hamiltonian system model of process systems, the above tools and techniques are then applied to process systems.
- *Simple process examples*
  A heat exchanger cell model and the model of the free mass convection network are used to illustrate the design of the nonlinear PD output feedback controller.
- *Stabilization of a simple unstable CSTR*
  Besides the stabilizing property of the nonlinear PD output feedback controller, the design parameters and the stability region of the controlled system are also investigated in the example of a simple unstable CSTR.
- *Hamiltonian control of a simple continuous fermenter*
  In this section, a complete case study is presented where various nonlinear PD output feedback controller structures are applied to the system, and their performance is compared.

12.1 Stabilization of Hamiltonian Systems

Let us consider a simple Hamiltonian system (see Definition 8.4.1 in Subsection 8.4.2) with an *equilibrium point*:

$$x_0 = (q_0, p_0)$$

It immediately follows from

$$\dot{q} = \frac{\partial H_0}{\partial p} = G(q)p$$

that $p_0 = 0$. Furthermore q_0 satisfies grad $V(q_0) = 0$, because it is an equilibrium point.

12.1.1 Asymptotic Stabilization of BIBO-stable Systems

First, we show that an equilibrium point of a simple Hamiltonian system can be stable without being asymptotically stable.

Lemma 12.1.1. *Let $(q_0, 0)$ be an equilibrium point of the simple Hamiltonian system given by Equations (8.32)–(8.33). Suppose that $V(q) - V(q_0)$ is a positive definite function in some neighborhood of q_0. Then the system for $u = 0$ is stable but not asymptotically stable.*

Proof. We will show that

$$\mathcal{L}(q, p) = H_0(q, p) - V(q_0) = \frac{1}{2}p^T G(q)p + V(q) - V(q_0)$$

is a Lyapunov function for a system with $u = 0$. Indeed,

$$\frac{d}{dt}\mathcal{L}(q, p) = \frac{d}{dt}H_0(q, p) = \sum_{i=1}^{n} \left(\frac{\partial H_0}{\partial q_i}\dot{q}_i + \frac{\partial H_0}{\partial p_i}\dot{p}_i \right)(q, p)$$

$$= \sum_{i=1}^{n} \left(\frac{\partial H_0}{\partial q_i}\frac{\partial H_0}{\partial p_i} - \frac{\partial H_0}{\partial p_i}\frac{\partial H_0}{\partial q_i} \right)(q, p) = 0 \qquad (12.1)$$

Note that we have used the definitions and the state-space model form in Equations (8.29)–(8.31) to derive the above result.

Furthermore, since $G(q) > 0$ and by assumption $V(q) - V(q_0)$ is positive definite, it follows that $\mathcal{L}(p, q)$ is positive in some neighborhood of $(q_0, 0)$. Since by Equation (12.1) $\frac{d}{dt}\mathcal{L}(q, p) = 0$, it also follows that $(q_0, 0)$ cannot be an asymptotically stable equilibrium.

Derivative Output Feedback. We can improve the situation, that is, to make this stable equilibrium point asymptotically stable by introducing a *derivative output feedback* to every input–output pair as follows:

$$u_i = -\dot{y}_i = -\frac{dH_i}{dt}, \quad i = 1, \dots, m \qquad (12.2)$$

which physically means adding damping to the system.

Indeed, with this feedback we obtain (again using Equations (8.29)–(8.31) with (12.1) and (12.2)):

$$\frac{d}{dt}\mathcal{L}(q,p) = \sum_{i=1}^{n}\left(\frac{\partial H_0}{\partial q_i}\dot{q}_i + \frac{\partial H_0}{\partial p_i}\dot{p}_i\right)$$

$$= \sum_{i=1}^{n}\left(\frac{\partial H_0}{\partial q_i}\frac{\partial H_0}{\partial p_i} - \frac{\partial H_0}{\partial p_i}\left(\frac{\partial H_0}{\partial q_i} + \sum_{j=1}^{m}\frac{\partial H_j}{\partial q_i}u_j\right)\right)$$

$$= \sum_{i=1}^{n}\sum_{j=1}^{m}\left(\dot{q}_i\frac{\partial H_j}{\partial q_i}\right)u_j$$

which finally yields

$$\frac{d}{dt}\mathcal{L}(q,p) = \frac{d}{dt}H_0(q,p) = -\sum_{i=1}^{m}\dot{y}_i^2 < 0 \tag{12.3}$$

12.1.2 Stabilization by Shaping the Potential Energy

The main assumption in Lemma 12.1.1 was the positive definiteness of the difference $V(q) - V(q_0)$ near an equilibrium point q_0. If this assumption does not hold, then let us apply a *linear proportional static output feedback*

$$u_i = -k_i y_i + v_i, \quad i = 1,\dots,m \tag{12.4}$$

with v_i being the new control inputs to the simple Hamiltonian system (8.32)–(8.33).

The Closed-loop Hamiltonian System. The resulting system is again a simple Hamiltonian system. In order to show this, we substitute the feedback (12.4) into the defining equations to obtain:

$$\overline{H}(q,p,v) = \frac{1}{2}p^T G(q)p + V(q) + \sum_{j=1}^{m}H_j^2(q)k_j - \sum_{j=1}^{m}H_j(q)v_j$$

with the internal Hamiltonian

$$\overline{H}_0(q,p) = \frac{1}{2}p^T G(q)p + \overline{V}(q) \tag{12.5}$$

where $\overline{V}(q)$ is the new potential energy

$$\overline{V}(q) = V(q) + \sum_{i=1}^{m}k_i y_i^2 \tag{12.6}$$

and with the state-space model:

$$\dot{q}_i = \frac{\partial \overline{H}_0}{\partial p_i}(q,p), \quad i = 1,\dots,n$$

$$\dot{p}_i = -\frac{\partial \overline{H}_0}{\partial q_i}(q,p) + \sum_{j=1}^{m}\frac{\partial H_j}{\partial q_i}(q)\,v_j, \quad i = 1,\dots,n$$

$$y_j = H_j(q), \quad j = 1,\dots,m$$

Equation (12.6) shows that we have added a positive term to the "old" potential energy.

By choosing sufficiently large feedback gains $k_i > 0$, we may shape the potential energy in such a way that it becomes positive definite.

12.1.3 The Nonlinear PD-controller for Hamiltonian Systems

If we want to stabilize a simple Hamiltonian system, then we should compose the two steps in the previous subsections:

- first, we make the system stable by suitable shaping the potential energy, and
- then asymptotically stabilize the system by adding damping to the new controls v_i.

Definition 12.1.1 (PD output feedback controller)
The following proportional-derivative (PD) output feedback controller

$$u_i = -k_i y_i - c_i \dot{y}_i, \quad k_i > 0 \,, \; c_i > 0, \quad i = 1, \ldots, m \tag{12.7}$$

is applied to simple Hamiltonian system models.

The results in Subsections 12.1.1 and 12.1.2 show that the PD output feedback controller above can locally asymptotically stabilize a simple Hamiltonian system with sufficiently large k_i. The additional freedom in the choice of the gain parameters (k_i, c_i, $i = 1, \ldots, m$) can be used to ensure a satisfactory transient behavior (analogously to the classical PD-controller case for a linear second-order system).

Properties of the PD Output Feedback Controller

1. It is important to note that the nonlinear PD output feedback relates "corresponding" $u_i - y_i$ pairs where y_i is the natural output to the input u_i. Analogously to the linear case, standard controllability and observability conditions should hold to ensure that we have "enough" inputs to achieve stabilization as $m \leq n$ in the general case.
2. Observe that the feedback (12.7) is only seemingly an output feedback because y_i is the "natural" and not the real output of the system. Taking into account the definition of the natural output of simple Hamiltonian systems in Equations (8.33) and (8.31), it turns out that y_i depends on the state variables in a nonlinear way, therefore the feedback (12.7) is in fact a nonlinear state feedback.

12.1.4 Comparison of the Nonlinear PD-controller and Feedback Linearization

If one compares the stabilization of a nonlinear system performed by feedback linearization or by the nonlinear PD-controller, the following points can be taken into account:

1. *Applicability range*
   The applicability range of the two controllers is much different with little overlap. Feedback linearization requires systems with relative degree $r = n$ or otherwise only approximative solutions may exist. The nonlinear PD-controller is only applicable to simple Hamiltonian systems but they include mechanical and process systems as well.
2. *Robustness*
   Feedback linearization requires the exact knowledge of the nonlinear system dynamics and it is known to be very sensitive with respect to mismodeling. The nonlinear PD-controller does only need the type of nonlinearity but it is not sensitive with respect to the parameters.
3. *Computational resources*
   The design of a nonlinear PD-controller does not require any computational efforts while feedback linearization is rather demanding from the viewpoint of the design computations.

12.2 Stabilization and Loop-shaping of Nonlinear Process Systems

We have already seen how to develop the Hamiltonian system models of process systems in Section 8.6, where the system variables and the elements of the system Hamiltonian have been introduced. This Hamiltonian system model serves as the basis of stabilization and loop-shaping of process systems.

12.2.1 Process Systems as Simple Hamiltonian Systems

The basis of the Hamiltonian description of a process system is the construction described in Section 8.6. It has been shown that a process system with the input variables (4.12) as the flow rates, the state variables (8.60) as the conserved extensive quantities in each balance volume, and the co-state variables (8.61) as the related thermodynamical potentials, enables us to construct a simple Hamiltonian system model with the Hamiltonian function (8.72) and with the underlying relationships (8.65), (8.68) and (8.62), (8.69).

The following special properties of the elements of the Hamiltonian description can be identified for process systems:

1. *Internal Hamiltonian*
   The internal Hamiltonian is related to the source term of the general conservation balances only, such that:

$$H_0(q,p) = V_Q(q), \quad \frac{\partial V_Q(q)}{\partial q} = -Q_\phi(q)$$

Therefore $H_0(q,p) = 0$ for process systems with no source.

2. *Coupling Hamiltonians*
   Recall, that the Hamiltonian description assumes that the input variables are the flow rates of the system, *i.e.* $u_j = v^{(j)}$. Then it follows from the general state equation (8.68) that

$$\frac{\partial H_j}{\partial q_i}(q) = (N_j Q^{-1} q)_{ij} + B_{conv,ij}$$

From this we see that the natural output is nonlinear and at most a quadratic function of the state variables q.

12.2.2 Stabilization

The stabilization of a Hamiltonian process system is then a simple application of the PD output feedback controller in Section 12.1.3, given in Equation (12.7).

From this we can see that a process system can be stabilized by the nonlinear PD output feedback controller in the form of (12.7) if the following conditions hold:

1. The system possesses a passive mass convection network (see in Subsection 4.2.4).
2. All of the flow rates are used for the feedback as input variables.

12.3 Simple Process Examples

This section illustrates the design of stabilization and loop-shaping controllers based on Hamiltonian system models using simple process examples introduced earlier in Chapter 4. A nonlinear PD output feedback controller is designed here for the bilinear heat exchanger cell and the free mass convection network.

12.3.1 Hamiltonian Control of the Heat Exchanger Cell

The basic Hamiltonian description of the nonlinear heat exchanger cell model has already been developed in Subsection 8.8.2 from its state-space model described in Section 4.4, where the state and co-state variables have been

identified and the internal Hamiltonian $H_0(q, p)$ has been derived in Equation
(8.85). We only need to identify the coupling Hamiltonians $H_1(q)$ and $H_2(q)$
from the co-state equation (8.83) for the input variables

$$u = [\ v_h - v_h^* \ , \ v_c - v_c^* \ ]^T$$

The coupling Hamiltonians can be reconstructed from the vector functions
$g_1(p)$ and $g_2(p)$ respectively, which are the gradient vectors of the corresponding coupling Hamiltonians:

$$\frac{\partial H_1}{\partial q} = g_1(q), \quad \frac{\partial H_2}{\partial q} = g_2(q)$$

By partial integration, we have already obtained in Equations (8.86) and
(8.87) that

$$H_1(q) = \frac{1}{2} c_{Ph} \left(T^*\right)^2 q_1^2 + c_{Ph} \overline{T}_{hi} q_1$$

$$H_2(q) = \frac{1}{2} c_{Pc} \left(T^*\right)^2 q_2^2 + c_{Pc} \overline{T}_{ci} q_2$$

From passivity analysis we know that the system is inherently passive but
that it has a pole at the stability boundary. Therefore we can perform stabilization by derivative feedback and loop-shaping by static feedback by using
two nonlinear PD output feedback controllers as follows:

(loop 1)

$$u_1 = -k_1 y_1 - c_1 \dot{y}_1, \quad k_1 > 0 \ , \ c_1 > 0 \tag{12.8}$$
$$y_1 = \frac{1}{2} c_P^{(h)} \left(T^*\right)^2 q_1^2 + c_P^{(h)} \overline{T}_{hi} q_1$$

(loop 2)

$$u_2 = -k_2 y_2 - c_2 \dot{y}_2, \quad k_2 > 0 \ , \ c_2 > 0 \tag{12.9}$$
$$y_2 = \frac{1}{2} c_P^{(h)} \left(T^*\right)^2 q_1^2 + c_P^{(h)} \overline{T}_{hi} q_1$$

12.3.2 Loop-shaping Control of the Free Mass Convection Network

Loop-shaping control of the free convection network described in Section
4.2.4 is developed here. The simple Hamiltonian system model is the basis
of the controller design which is described in Subsection 8.8.3 using the LTI
state-space model of the free mass convection network in Equation (4.22).

The state and co-state variables, which are the volume and the hydrostatic
pressure in each balance volume,

$$q = [\ \rho g \overline{h}^{(1)} \ , \dots, \ \rho g \overline{h}^{(C)} \ ]^T$$

$$p = [\ \overline{V}^{(1)}\ , \ldots,\ \overline{V}^{(C)}\ ]^T$$

have been identified and the internal Hamiltonian $H_0(q, p)$ has been derived in Equation (8.89).

The coupling Hamiltonians can be derived from the co-state equations (8.88), taking into account that the input functions are the gradient vectors of the corresponding coupling Hamiltonians. We observe that an input variable $u_j = v_{IN}^{(j)}$ acts only on its state variable $q_j = \rho g \overline{h}^{(j)}$, therefore

$$H_j(q_j) = -\frac{1}{\rho} q_j = -\chi_j \overline{h^{(j)}}$$

From passivity analysis we know that the system is inherently passive with no poles at the stability boundary. Therefore we need not perform stabilization by a derivative output feedback but can perform loop-shaping by static feedback by using C pieces of P-controllers as follows:

(loop j)

$$u_j = -k_j y_j, \quad k_j > 0, \quad y_j = \chi_j \overline{h^{(j)}}\ ,\ \chi_j > 0 \tag{12.10}$$

The above separate static linear output feedback controllers perform pole-placement in a similar way to what we have seen in the case of pole-placement controllers (see Section 9.2) but now in the context of a MIMO setting as a non-optimal LQR (see Section 9.3).

12.4 Stabilization of a Simple Unstable CSTR

Consider the same unstable CSTR as in Section 8.8.4. There we described an isotherm CSTR with fixed mass hold-up m and constant physico-chemical properties. A second-order

$$2A + S \rightarrow T + 3A$$

autocatalytic reaction takes place in the reactor where the substrate S is present in great excess. Assume that the inlet concentration of component A (c_{Ain}) is constant and that the inlet mass flow rate v is used as an input variable. The Hamiltonian description of the system has been given around its steady-state point determined as the setpoint for passivation and loop-shaping in Section 8.8.4.

Now we develop a loop-shaping controller based on the Hamiltonian system model [30].

12.4.1 System Parameters and Open-loop Response

Let us introduce the normalized concentration variables $\bar{c}_A = c_A - c_A^*$ and $\bar{c}_{Ain} = c_{Ain} - c_A^*$. The conservation balance equation (4.70) then takes the form

$$\frac{d\bar{c}_A}{dt} = k \cdot \bar{c}_A^2 + (2k \cdot c_A^* - \frac{v^*}{m}) \cdot \bar{c}_A + \frac{\bar{c}_{Ain} - \bar{c}_A}{m} \cdot u \tag{12.11}$$

The parameter values used in the simulations are shown in Table 12.1.

Table 12.1. Parameter values of the simulated CSTR

Variable	Value	Unit
m	1800	kg
k	$5 \cdot 10^{-4}$	$\frac{m^3}{\text{kmol;s}}$
c_{Ain}	0.4	$\frac{\text{kmol}}{m^3}$
c_A^*	2.3	$\frac{\text{kmol}}{m^3}$
v^*	2.5058	$\frac{\text{kg}}{s}$
k_c	10	-

There were two initial concentration values $(c_A(0))$ given for the simulations:

- 1 $\frac{\text{kmol}}{m^3}$ and
- 2.8 $\frac{\text{kmol}}{m^3}$

respectively.

It is easily seen from the data that c_A^* is an unstable equilibrium for the system as it is illustrated in both of the sub-figures in Figure 12.1.

12.4.2 Nonlinear Proportional Feedback Controller

Let us apply the following nonlinear proportional output feedback controller

$$u = k_c y_1 + w = k_c \cdot (\frac{1}{2}\bar{c}_A^2 - \bar{c}_{Ain} \cdot \bar{c}_A) + w \tag{12.12}$$

where k_c is an appropriately chosen controller gain and w is the new reference signal. The new reference w was set to 0 for the simulations. The chosen value of the controller gain is shown in Table 12.1 together with the system parameters.

The closed-loop simulation results in Figure 12.2 show that the proposed control method indeed stabilizes the equilibrium $c_A^* = 2.3 \frac{\text{kmol}}{m^3}$. Here again, the simulation was performed using two different initial conditions as above as shown in the two sub-figures of Figure 12.2.

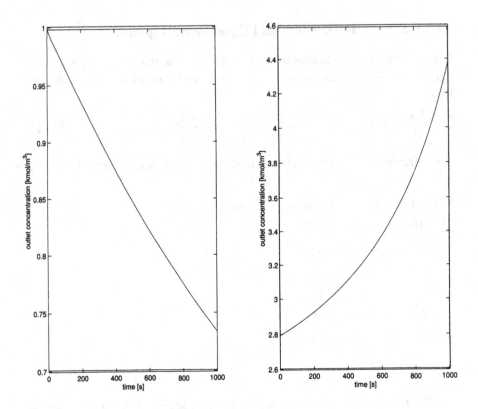

Figure 12.1. Open-loop simulation results

12.4.3 Stability Region

It is an important question for a nonlinear controller to determine its stability region as a function of the state variables with its parameter(s) fixed. For this very simple case this problem can be solved analytically.

Let us consider the nonlinear proportional feedback controller above with its gain fixed at $k_c = 10$. In fact, it is easy to show that the resulting closed-loop system with the parameters described above is passive with respect to the supply rate $w \cdot y$ if

$$\bar{c}_A > -1.9088 \ \frac{\text{kmol}}{\text{m}^3} \quad i.e. \quad c_A > 0.3912$$

where $y = \bar{c}_A$.

In order to show this, let us take the simple storage function $V(\bar{c}_A) = \frac{1}{2}\bar{c}_A^2$. It can be calculated that

$$\frac{\partial V}{\partial \bar{c}_A} \cdot \dot{\bar{c}}_A = \frac{\partial V}{\partial \bar{c}_A} \cdot f(\bar{c}_A) \leq 0 \quad \text{if } \bar{c}_A > -1.9088 \tag{12.13}$$

and equality holds only if $\bar{c}_A = 0$. Since $g(\bar{c}_A) = 1$ in the closed-loop state-space model we can deduce that

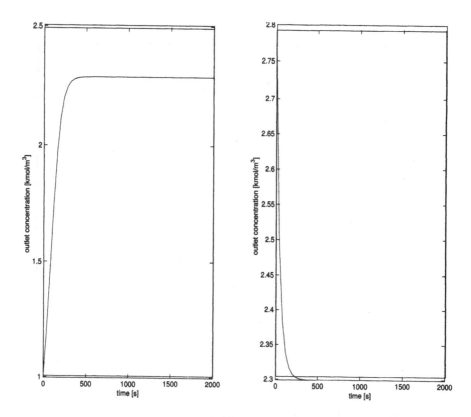

Figure 12.2. Closed-loop simulation results

$$y = L_g V(\bar{c}_A) = \frac{\partial V}{\partial \bar{c}_A} \cdot g(\bar{c}_A) = \bar{c}_A \tag{12.14}$$

where L_g is the Lie-derivative of the simple storage function with respect to the function g. Therefore it follows that the closed-loop system is passive in the given interval.

The time derivative of the storage function (as a function of $\bar{c}_A$) is depicted in Figure 12.3.

12.5 Hamiltonian Control of a Simple Continuous Fermenter

The simple continuous fermenter is used as a case study for designing a nonlinear PD output feedback controller for a process system. The input-affine nonlinear state-space model of the fermenter is developed in Section 4.5.3 and it is given in Equations (4.78)–(4.79). The variables and parameters of the fermentation process model are collected in Table 4.1.

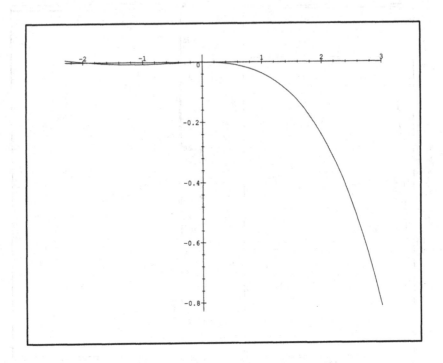

Figure 12.3. Time derivative of the storage function as a function of $\bar{c}_A$

12.5.1 Hamiltonian Model of the Fermentation Process

Based on the input-affine state-space model, it is fairly easy to obtain the Hamiltonian system model of the fermentation process. For this, let us introduce the following variables:

$$p = \begin{bmatrix} p_1 \\ p_2 \end{bmatrix} = \begin{bmatrix} m_X - m_{X_0} \\ m_S - m_{S_0} \end{bmatrix} = \begin{bmatrix} \bar{m}_X \\ \bar{m}_S \end{bmatrix} \tag{12.15}$$

$$q = \begin{bmatrix} q_1 \\ q_2 \end{bmatrix} = -\begin{bmatrix} \bar{X} \\ \bar{S} \end{bmatrix} \tag{12.16}$$

where $p = -Vq$ denotes the centered component masses. The Hamiltonian form of the model is written as

$$\dot{p}_1 = -V(X_0 - q_1)\mu(S_0 - q_2) + (X_0 - q_1)F_0 + (X_0 - q_1)\bar{F}$$
$$\dot{p}_2 = \frac{V(X_0 - q_1)\mu(S_0 - q_2)}{Y} - (S_F - (S_0 - q_2))F_0 - (S_F - (S_0 - q_2))\bar{F}$$

Then the coupling Hamiltonian (natural output) used for feedback is calculated by simple integration. Since

$$\frac{\partial H_1}{\partial q_1} = X_0 - q_1, \quad \frac{\partial H_1}{\partial q_2} = -(S_F - (S_0 - q_2)) \tag{12.17}$$

the natural output of the system is written as

$$y_1 = H_1(q) = X_0 q_1 - \frac{1}{2}q_1^2 - (S_F q_2 - (S_0 q_2 - \frac{1}{2}q_2^2)) \tag{12.18}$$

Note that the above natural output for feedback defines a *nonlinear static full state feedback which is a quadratic function of both state variables*.

12.5.2 Full State Feedback Using the Whole Natural Output

In this case, the feedback consists of the entire calculated natural output

$$\bar{F} = -k_1 y_1 = -k_1 (X_0 q_1 - \frac{1}{2}q_1^2 - S_F q_2 + S_0 q_2 - \frac{1}{2}q_2^2) \tag{12.19}$$

where k_1 is an appropriately chosen positive constant. Again, we emphasize that both of the biomass and the substrate concentration is needed to realize the nonlinear static feedback above. This calls for the application of a possibly nonlinear state filter to determine the biomass concentration X which is hardly measurable.

The simulation results corresponding to $k_1 = 0.5$ are visible in Figure 12.4.

12.5.3 State Feedback Using Only a Part of the Natural Output

It is known from the theory of Hamiltonian systems that the definiteness of the time derivative of the storage function (*i.e.* the stability of the closed-loop system) may be influenced by using only some part of the natural output for feedback. Since the biomass concentration is quite difficult to measure in practice, we choose the substrate concentration for feedback, *i.e.*

$$\bar{F} = -k_2 \cdot -(S_F q_2 - (S_0 q_2 - \frac{1}{2}q_2^2)) \tag{12.20}$$

The simulation results for $k_2 = 1$ can be seen in Figure 12.5. As can be seen, the application of only one state variable is enough to stabilize the system in this case. Moreover, the quality of the control remains much the same as compared to the full state feedback case seen in Figure 12.4.

12.5.4 Controller Tuning and Stability Analysis of the Closed-loop System

A simple graphical method can be used for tuning the Hamiltonian controllers and investigating the stability region around a given operating point. The

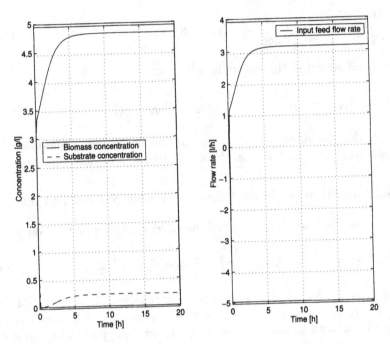

Figure 12.4. Typical result of Hamiltonian control using the entire natural output: state variables and input

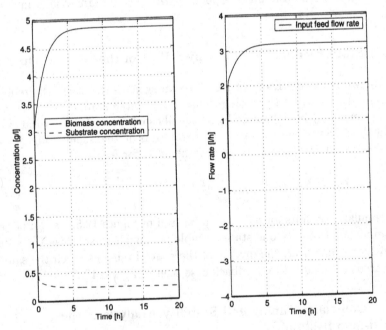

Figure 12.5. Typical result of Hamiltonian control using the substrate concentration only: state variables and input

method is applicable for systems where only one input variable with a scalar feedback gain k is applied.

For the purpose of the analysis, the state-space form of the closed-loop system will be denoted by

$$\dot{x} = \bar{f}(x) \tag{12.21}$$

where

$$\bar{f}(x) = f(x) + g(x)(-k_1 \cdot y_1(x)) \tag{12.22}$$

First, we construct a simple quadratic storage function

$$V(x) = \frac{1}{2}x^T x \tag{12.23}$$

and examine the sign of its time derivative around the desired operating point. This storage function is clearly positive definite and it is a candidate for being a Lyapunov function of the system.

The time derivative of the storage function is given by

$$\dot{V}(x) = \frac{d}{dt}V(x) = x^T \cdot \dot{x} = x^T \bar{f}(x) \tag{12.24}$$

where $\bar{f}$ is the right-hand side of the state-space model of the closed-loop system, which depends on the feedback gain k_1. Then the idea is to *choose the feedback gain in such a way that the above time derivative is negative definite over a wide operating region in the state-space.*

Hamiltonian Control with Full State Feedback. The time derivatives of the quadratic storage function as functions of the state variables X and S corresponding to the feedback gain $k_1 = 0.5$ and $k_1 = 5$ are shown in Figures 12.6. (a) and (b) respectively. It can be easily calculated that the desired operating point $(x_0 = [0,0]^T)$ is a local maximum of $\dot{V}$ at $\dot{V}(x_0) = 0$. Therefore the proposed full state Hamiltonian feedback locally asymptotically stabilizes the desired operating point. If we have a look at Figure 12.6 again, we can see that the stability region is fairly wide around the operating point in both cases, since the sign of $\dot{V}$ is negative everywhere except the close neighborhood of the singularity points $[X = 0, \ S = S_F]^T$ and $[X = 0, \ S \neq S_F]^T$.

Hamiltonian Control with Partial State Feedback. If we apply the same storage function for stability analysis as in the previous case, we can again plot $\dot{V}$ as a function of the biomass and substrate concentrations. This can be seen in Figures 12.7. (a) and (b) corresponding to the feedback gains $k_1 = 0.5$ and $k_1 = 5$ respectively.

As can be seen, the system is again stable in a reasonable wide neighborhood of the operating point in these cases. This is in good agreement with what we expected from the simulation results, *i.e.* that using only the substrate concentration for nonlinear feedback is sufficient for stabilizing the system.

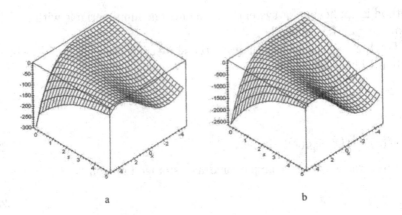

Figure 12.6. Time derivative of the quadratic Lyapunov function as a function of X and S, Hamiltonian full state feedback, (a) $k_1 = 0.5$, (b) $k_1 = 5$

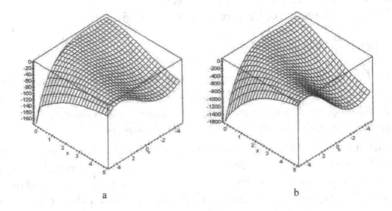

Figure 12.7. Time derivative of the quadratic Lyapunov function as a function of X and S, Hamiltonian partial state feedback, (a) $k_1 = 5$, (b) $k_1 = 5$

12.5.5 Discussion

From the above cases, we can clearly see that the stabilizability region and the value of the feedback gain are closely related, therefore they should be investigated/designed jointly. The above proposed graphical method can be used for *simultaneous feedback control gain selection and stabilization region determination*.

12.6 Further Reading

The traditional and still main application area of Hamiltonian system models and their physics-based control is in mechatronics (robots): see [65] for a simple introductory paper. A good early survey of control based on Hamiltonian system models in mechatronics is found in [52], while a recent paper on port-controlled Hamiltonian system is [51].

Process Systems. There is emerging literature in the field of making connections between thermodynamics and the theory of Hamiltonian systems for process control purposes.

The principles of constructing a Hamiltonian control of process systems can be found in [30].

A survey of related papers is offered by Ydstie [83].

12.7 Summary

A powerful robust method for stabilization and loop-shaping is presented in this chapter for nonlinear input-affine systems which enable a Hamiltonian system model. The stabilizing and loop-shaping controller is a proportional-derivative (PD) output feedback controller, which requires to feed back the artificial output related to each input. This controller in fact is a nonlinear static feedback controller, because both the artificial output and its time derivative are nonlinear static functions of the state variables.

Process systems of different complexity, a heat exchanger cell, free mass convection networks, a simple unstable CSTR and a simple continuous fermenter are used to demonstrate the use of the proposed PD output feedback controller on process systems.

12.8 Questions and Application Exercises

Exercise 12.8.1. Give the design parameters of the nonlinear PD output feedback controller design. Comment on the effect of these parameters on the properties of the closed-loop system.

Exercise 12.8.2. What is the role of the proportional and the derivative part of the nonlinear PD output feedback controller? Give process system examples, when one of the parts is not needed.

Exercise 12.8.3. Compare passivation by static feedback design described in Chapter 11 and stabilization and loop-shaping based on Hamiltonian system models. What are the similarities and the differences?

Exercise 12.8.4. Characterize the nonlinearities that can occur in the nonlinear PD output feedback controller when controlling process systems. What is the physical reason of this nonlinearity type?

Exercise 12.8.5. Compare the properties of the direct passivation controller presented in Section 11.3 and the Hamiltonian control of the same system in Section 12.5. Comment on the tuning parameters, computational requirements and on the stability region.

Exercise 12.8.6. Construct a nonlinear PD output feedback controller for the ideal spring model developed in Example 8.3.1.

Exercise 12.8.7. Consider a simple CSTR model developed in Subsection 4.5.1 but with the reaction rate expression

$$r = k \cdot m \cdot c_A^4$$

Construct a nonlinear PD output feedback controller for this CSTR following the way for the stabilization of the simple unstable CSTR described in Subsection 12.4.

A. Mathematical preliminaries

A.1 Vector and Signal Norms

If we have objects of non-scalar nature, for example vectors, matrices, functions or signals, we measure their magnitude by using *norms*. Norms are the extensions of a length of a vector applied to other objects of non-scalar nature forming a so called *vector space*.

Definition A.1.1 (Vector space)
The set of elements X with the operation addition and multiplication by scalar is called a vector space if

- *if $x_1, x_2 \in X$ then $x^* = (x_1 + x_2) \in X$*
- *if $x_1 \in X$ then $x^* = ax_1 \in X$ for any $a \in \mathbb{R}$*
- *there is a zero element $x_0 \in X$ for which $x_1 + x_0 = x_1$ holds for every $x_1 \in X$*

Definition A.1.2 (Norm on vector space)
A scalar-valued function $\rho(.) : X \rightarrow \mathbb{R}$ is a norm on a vector space X if

- *$\rho(x) \geq 0$, $\forall x \in X$*
- *$\rho(x) = 0 \Leftrightarrow x = 0 \in X$*
- *$\rho(x + y) \leq \rho(x) + \rho(y)$ (triangular inequality)*

A.1.1 Vector Norms

n-dimensional vectors with real-valued entries $x = [x_1, x_2, \cdots, x_n]$ form a vector space called $\mathbb{R}^n$. There are different known and useful norms defined over the vector space $\mathbb{R}^n$ which are as follows.

Definition A.1.3 (Vector norms)
1. p-norm

$$\|x\|_p = \left(\sum_{i=1}^{n} |x_i|^p \right)^{1/p} , \quad p = 1, 2, \ldots \tag{A.1}$$

2. Euclidean norm or 2-norm $(p = 2)$

$$||x||_2 = \sqrt{\sum_{i=1}^{n} x_i^2} \tag{A.2}$$

3. maximum norm or ∞-norm $(p \to \infty)$

$$||x||_\infty = \max_i(|x_i|) \tag{A.3}$$

A.1.2 Signal Norms

Systems are usually described by differential equations which contain *signals* f from a signal space $\mathcal{L}$ (compare with Section 2.1) A signal my be *scalar-valued or vector-valued*, i.e. :

$$f(t) \in \mathcal{R} \quad , \quad F(t) \in \mathcal{R}^\mu$$

for any fixed time instant t. We shall use two types of *norms of a signal*: the *infinity norm* and the *2-norm*.

Definition A.1.4 (Signal norms)
- p-norm

$$||f||_p = \left(\int_0^\infty |f(t)|^p \right)^{1/p} \tag{A.4}$$

- 2-norm $(p = 2)$

$$||f||_2^2 = \int_{-\infty}^\infty f^2(t) \quad , \quad ||F||_2^2 = \int_{-\infty}^\infty ||F(t)||_2^2 \tag{A.5}$$

- infinity norm $(p \to \infty)$

$$||f||_\infty = \sup_t |f(t)| \quad , \quad ||F||_\infty = \sup_t ||F(k)||_\infty \tag{A.6}$$

A.2 Matrix and Operator Norms

Definition A.2.1 (Induced operator norm)
The induced norm of an operator q on the vector space X induced by a norm $||.||$ on the same space is defined as

$$||q|| = \sup_{||x||=1} \frac{||q(x)||}{||x||} \tag{A.7}$$

Matrices can be seen as linear operators transforming vectors from a vector space like $\mathbb{R}^n$ to another vector space which can be the same or a different one like $\mathbb{R}^m$. For rectangular matrices when $m \neq n$ and $T \in \mathbb{R}^{m \times n}$ we have different vector spaces and for square matrices when $n = m$ they are the same vector space.

The "magnitude" of a matrix is also characterized by its norm. Similarly to the case of vectors and matrices there are various norms applicable for matrices. The most important ones are the so called *induced norms* where the matrix norm is derived using an already defined vector norm.

Definition A.2.2 (Matrix norm, induced)
The induced N-norm *of a square matrix $T \in \mathcal{R}^{n \times n}$ is defined as:*

$$\|T\|_N = \sup_{x \in \mathbb{R}^n} \frac{\|Tx\|_N}{\|x\|_N} \tag{A.8}$$

where $\|.\|_N$ is a vector norm.

A.3 Lie-derivative, Lie-product

The notion of Lie-derivative and Lie-product play central role in the geometric approach of nonlinear system analysis and control.

A.3.1 Lie-derivative

Definition A.3.1 (Lie-derivative)
Let us given a nonlinear scalar-valued function $\lambda \in \mathbb{R}^n \mapsto \mathbb{R}$ and a vector field $f \in \mathbb{R}^n \mapsto \mathbb{R}^n$ on a common domain $U = dom(\lambda) = dom(f) \subseteq \mathbb{R}^n$ open and let λ be continuously differentiable, i.e. $\lambda \in C^1$ on U
   The derivative of λ along f is defined as

$$L_f\lambda(x) = \frac{\partial \lambda(x)}{\partial x} f(x) = \sum_{i=1}^{n} \frac{\partial \lambda(x)}{\partial x_i} f_i(x) = < d\lambda(x), f(x) > \tag{A.9}$$

Observe that the vector field f in any point defines a direction along which we compute the derivative (gradient) of λ.

It is important to note that $L_f\lambda \in \mathbb{R}^n \mapsto \mathbb{R}$. The repeated Lie derivation is defined recursively:

$$L_f^k\lambda(x) = \frac{\partial(L_f^{k-1}\lambda(x))}{\partial x} f(x) \tag{A.10}$$

with $L_f^0\lambda = \lambda$.

Example A.3.1 (Simple Lie-derivatives)

Two simple examples illustrate the above definition.

1. one-dimensional example, i.e. $x \in \mathbb{R}$

$$\lambda(x) = x^2 \quad , \quad f(x) = \sqrt{x}$$

Now we can compute both $L_f\lambda$ and $L_\lambda f$ as follows:

$$L_f\lambda(x) = 2x \cdot \sqrt{x} \quad , \quad L_\lambda f(x) = \frac{1}{2\sqrt{x}} \cdot x^2$$

2. two-dimensional example, $x \in \mathbb{R}^2$

$$\lambda(x_1, x_2) = x_1^2, \quad f(x_1, x_2) = f(x) = \begin{bmatrix} x_1^2 + x_2^2 \\ x_1^3 + x_2^3 \end{bmatrix}$$

and

$$L_f\lambda(x_1, x_2) = 2x_1(x_1^2 + x_2^2).$$

A.3.2 Lie-product

Lie-product is defined for two vector fields $f, g \in \mathbb{R}^n \mapsto \mathbb{R}^n$ defined on a joint open domain $U = dom(f) = dom(g) \subseteq \mathbb{R}^n$ where both f and g are continuously differentiable (C^1 on U)

Definition A.3.2 (Lie-product (Lie-bracket))
The Lie-product of f and g is defined as:

$$[f, g](x) = \frac{\partial g(x)}{\partial x} f(x) - \frac{\partial f(x)}{\partial x} g(x) \tag{A.11}$$

where

$$\frac{\partial g(x)}{\partial x} = \begin{bmatrix} \frac{\partial g_1(x)}{\partial x_1} & \frac{\partial g_1(x)}{\partial x_2} & \cdots & \frac{\partial g_1(x)}{\partial x_n} \\ \frac{\partial g_2(x)}{\partial x_1} & \frac{\partial g_2(x)}{\partial x_2} & \cdots & \frac{\partial g_2(x)}{\partial x_n} \\ \vdots & \vdots & \ddots & \vdots \\ \frac{\partial g_n(x)}{\partial x_1} & \frac{\partial g_n(x)}{\partial x_2} & \cdots & \frac{\partial g_n(x)}{\partial x_n} \end{bmatrix} \tag{A.12}$$

is the Jacobian matrix of the function g with respect to its independent variable x.

Naturally, $[f, g](x) \in \mathbb{R}^n$ which enables to apply the Lie-bracket operation repeatedly.

Important Properties of Lie-bracket.

- **Bilinearity over $\mathbb{R}$**

  $f_1, f_2, g_1, g_2 \in \mathbb{R}^n \mapsto \mathbb{R}^n$, $r_1, r_2 \in \mathbb{R}$

$$[r_1 f_1 + r_2 f_2, g_1] = r_1 [f_1, g_1] + r_2 [f_2, g_2] \qquad \text{(A.13)}$$
$$[f_1, r_1 g_1 + r_2 g_2] = r_1 [f_1, g_1] + r_2 [f_1, g_2] \qquad \text{(A.14)}$$

- **Skew commutativity** $f, g \in \mathbb{R}^2 \mapsto \mathbb{R}^n$

$$[f, g] = -[g, f] \qquad \text{(A.15)}$$

- **Jacobi identity**

$$[f, [g, p]] + [g, [p, f]] + [p, [f, g]] = 0 \qquad \text{(A.16)}$$

The following simple examples show further important properties of Lie-bracket which are useful in process system applications.

Example A.3.2 (Simple Lie-products)

The following simple two-dimensional ($x \in \mathbb{R}^2$) examples illustrate the definition of Lie-products.

1. Two general vector fields f and g

$$f(x) = \begin{bmatrix} x_1^2 + x_2^2 \\ x_1^3 + x_2^3 \end{bmatrix} \quad , \quad g(x) = \begin{bmatrix} -x_1 \\ x_2 \end{bmatrix}$$

   gives the following Lie-product

$$[f, g](x) = \begin{bmatrix} -1 & 0 \\ 0 & 1 \end{bmatrix} \cdot \begin{bmatrix} x_1^2 + x_2^2 \\ x_1^3 + x_2^3 \end{bmatrix} - \begin{bmatrix} 2x_1 & 2x_2 \\ 3x_1^2 & 3x_2^2 \end{bmatrix} \cdot \begin{bmatrix} -x_1 \\ x_2 \end{bmatrix}$$

2. A constant vector field $\overline{f}$ with a general vector field g

$$\overline{f}(x) = \begin{bmatrix} 3 \\ 2 \end{bmatrix} \quad , \quad g(x) = \begin{bmatrix} -x_1 \\ x_2 \end{bmatrix}$$

   gives the following Lie-product

$$[\overline{f}, g](x) = \begin{bmatrix} -3 \\ 2 \end{bmatrix}$$

   where only the first term $\frac{\partial g}{\partial x} f$ is present.

---

3. Two constant vector fields $\bar{f}$ and $\bar{g}$

$$\bar{f}(x) = \begin{bmatrix} 3 \\ 2 \end{bmatrix} \quad , \quad \bar{g}(x) = \begin{bmatrix} 4 \\ 5 \end{bmatrix}$$

gives the following Lie-product

$$[\bar{f}, \bar{g}](x) = \begin{bmatrix} 0 \\ 0 \end{bmatrix}$$

because both $\frac{\partial g}{\partial x}$ and $\frac{\partial f}{\partial x}$ are zero matrices.

A.4 Distributions, Co-distributions

Distributions and co-distributions are important basic notions in the geometric approach to analyzing dynamic properties of nonlinear systems, such as controllability and observability.

A.4.1 Distributions

Definition A.4.1 (Distribution)
Consider a set of functions $f_1, \ldots, f_d \in \mathbb{R}^n \to \mathbb{R}^n$ with joint domain U in $\mathbb{R}^n$:

$$U = \text{dom}(f_1) =, \ldots, = \text{dom}(f_d) \subseteq \mathbb{R}^n$$

The distribution Δ spanned by the functions $f_1, \ldots, f_d$

$$\Delta(x) = \text{span}\{f_1(x), \ldots, f_d(x)\} \tag{A.17}$$

assigns a vector space to each point x_0 of U.
In notation we omit the independent variables and write:

$$\Delta = \text{span}\{f_1, \ldots, f_d\} \tag{A.18}$$

Definition A.4.2 (Dimension of a distribution)
The dimension of a distribution Δ at x is the dimension of the vector space spanned by Δ in x.

Operations on Distributions. The following set-like operations are defined on distributions:

$$(\Delta_1 + \Delta_2)(x) = \Delta_1(x) + \Delta_2(x) \tag{A.19}$$
$$(\Delta_1 \cap \Delta_2)(x) = \Delta_1(x) \cap \Delta_2(x) \tag{A.20}$$

Obviously, if the distributions Δ_1 and Δ_2 are spanned by the functions $(f_1, \ldots, f_n)$ and $(g_1, \ldots, g_m)$ respectively, then the distribution $(\Delta_1 + \Delta_2)$ is spanned by

$$(f_1, \ldots, f_n, g_1, \ldots, g_m)$$

.

Distributions and their Properties. There are some notions and special distribution types which are important from practical point of view.

- *Nonsingular distribution Δ defined on U*
  Δ is a nonsingular distribution if $\exists d \in \mathbb{N}$ such that

$$\dim(\Delta(x)) = d \quad \forall x \in U \tag{A.21}$$

- *Regular point of a distribution Δ*
  x_0 is a regular point of distribution Δ if there exists a neighborhood U^0 of x^0 such that Δ is nonsingular on U^0.
- *Point of singularity*
  x_S is a point of singularity if it is not a regular point.
- *f belongs to the distribution Δ $(f \in \Delta)$ when*

$$f(x) \in \Delta(x) \quad \forall x \tag{A.22}$$

- *Distribution Δ_1 contains a distribution Δ_2 $(\Delta_1 \supset \Delta_2)$ if*

$$\Delta_1(x) \supset \Delta_2(x) \quad \forall x \tag{A.23}$$

- *Distribution Δ is involutive*
  A distribution Δ is called involutive if

$$\tau_1 \in \Delta, \tau_2 \in \Delta \Rightarrow [\tau_1, \tau_2] \in \Delta \tag{A.24}$$

- *Distribution Δ is invariant under the vector field f when*

$$\tau \in \Delta \Rightarrow [f, \tau] \in \Delta \tag{A.25}$$

The following simple examples show how to work with distributions.

Example A.4.1 (Simple distribution)

Consider a set of functions $f_i : \mathbb{R}^n \to \mathbb{R}^n$ such that

$$f_i = \begin{bmatrix} 0 \\ \cdots \\ 0 \\ x_i \\ 0 \\ \cdots \\ 0 \end{bmatrix}$$

where only the ith entry is non-zero for $i = 1, \ldots, n$.
The dimension of the distribution

$$\Delta(x) = \text{span}\{f_1, \ \ldots, \ f_n\}$$

depends on the point x as follows.

- $\dim(\Delta(x)) = n$ when $x_i \neq 0$, $i = 1, \ldots, n$
- $\dim(\Delta(x)) = d < n$ when at least one of the entries of x is zero, i.e. $\exists i, \ x_i = 0$
- $\dim(\Delta(x)) = 0$ if $x = 0$, i.e. in the origin of the space $\mathcal{X}$

Therefore Δ is nonsingular everywhere except

$$\mathcal{D}_{sing} = \{x \mid \exists i, \ x_i = 0, \ i = 1, \ldots, n\}$$

The above set $\mathcal{D}_{sing}$ contains all singular points of distribution Δ.

A.4.2 Co-distributions

Co-distributions are defined using the notions of dual space and co-vector fields which are as follows.

Definition A.4.3 (Dual space of a vector space)
Dual space V^* of a vector space $V \subset \mathbb{R}^n$ is the set of all linear real-valued functions defined on V. Formally defined as

$$f(x) = f(x_1, x_2, \ldots, x_n) = a_1 x_1 + a_2 x_2 + \cdots + a_n x_n \tag{A.26}$$
$$a_i \in \mathbb{R}, i = 1, \ldots, n$$

i.e. $f(x) = ax$ where

$$a = [a_1 \; a_2 \; \dots \; a_n] \qquad x = \begin{bmatrix} x_1 \\ x_2 \\ \vdots \\ x_n \end{bmatrix} \qquad \text{(A.27)}$$

Note that f is given by the row vector a.

Definition A.4.4 (Co-vector field)
A mapping from $\mathbb{R}^{n \times 1}$ to $\mathbb{R}^{1 \times n}$ is called a co-vector field. It can be represented row vector valued function.

$$f(x) = f(x_1, x_2, \dots, x_n) = [f_1(x) \; f_2(x) \; \dots \; f_n(x)] \qquad \text{(A.28)}$$

Example A.4.2 (Gradient, a special co-vector field)

Let us define a co-vector field $d\lambda$ associated to a vector field $\lambda \in \mathbb{R}^n \to \mathbb{R}$ as follows:

$$d\lambda(x) = \left[\frac{\partial \lambda(x)}{\partial x_1} \; \frac{\partial \lambda(x)}{\partial x_2} \; \dots \; \frac{\partial \lambda(x)}{\partial x_n} \right] \qquad \text{(A.29)}$$

$d\lambda$ is called the gradient of λ.

Definition A.4.5 (Co-distribution)
Let $\omega_1, \dots, \omega_n$ be smooth co-vector fields. Then Ω is a co-distribution spanned by the co-vectors:

$$\Omega(x) = \text{span}\{\omega_1(x), \dots, \omega_d(x)\} \qquad \text{(A.30)}$$

At any point x_0 co-distributions are subspaces of $(\mathbb{R}^n)^*$.
    We usually omit the argument of the co-vector fields and the co-distribution and write:

$$\Omega = \text{span}\{\omega_1, \dots, \omega_d\} \qquad \text{(A.31)}$$

Operations on Co-distributions and their Properties. Operations, such as *addition, intersection, inclusion* are defined in an analogue way to that of distributions. Likewise, the notion of the dimension of a co-distribution at a point, regular point, point of singularity are applied to co-distributions in an analogous way.

Special Co-distributions and their Properties.

- *Annihilator of a distribution Δ $(\Delta^\perp)$*
  The set of all co-vectors which annihilates all vectors in $\Delta(x)$.

$$\Delta^\perp(x) = \{w^* \in (\mathbb{R}^n)^* | <w^*, v> = 0 \ \forall v \in \Delta(x)\}. \tag{A.32}$$

is called the annihilator of the distribution Δ.
The annihilator of a distribution is a co-distribution. The annihilator of a smooth distribution is not necessarily smooth.

- *Annihilator of a co-distribution Ω $(\Omega^\perp)$*

$$\Omega^\perp(x) = \{v \in \mathbb{R}^n | <w^*, v> = 0 \ \forall w^* \in \Omega(x)\}. \tag{A.33}$$

is the annihilator of a co-distribution Ω.
The annihilator of a co-distribution is a distribution.

- *Co-distribution invariant under the vector field f:*
  The co-distribution Ω is invariant under the vector field f if and only if

$$\omega \in \Omega \ \Rightarrow \ L_f \omega \in \Omega \tag{A.34}$$

where $(L_f \omega)(x) = \omega(x) \frac{\partial f(x)}{\partial x}$.

- *Sum of dimensions of a distribution and its annihilator*

$$dim(\Delta) + dim(\Delta^\perp) = n \tag{A.35}$$

- *Inclusion properties*

$$\Delta_1 \supset \Delta_2 \ \Longleftrightarrow \ \Delta_1^\perp \subset \Delta_2^\perp \tag{A.36}$$

- *Annihilator of an intersection*

$$(\Delta_1 \cap \Delta_2)^\perp = \Delta_1^\perp + \Delta_2^\perp \tag{A.37}$$

- *Compatibility of the dimension of a distribution*
  If a distribution Δ is spanned by the columns of a matrix F, the dimension of Δ at a point x^0 is equal to the rank of $F(x^0)$. If the entries of F are smooth functions of x then the annihilator of Δ is identified at each $x \in U$ by the set of row vectors w^* satisfying the condition $w^* F(x) = 0$.

- *Compatibility of the dimension of a co-distribution*
  If a co-distribution Ω is spanned by the rows of a matrix W, whose entries are smooth functions of x, its annihilator is identified at each x by the set of vectors v satisfying $W(x) = 0$, i.e.

$$\Omega^\perp(x) = ker(W(x)) .$$

- *Compatibility of the invariant property of annihilators*
  If a smooth distribution Δ is invariant under the vector field f, then the co-distribution $\Omega = \Delta^\perp$ is also invariant under f.
  If a smooth co-distribution Ω is invariant under the vector field f, then the distribution $\Delta = \Omega^\perp$ is also invariant under f.

- *Condition of involutivity*
  A smooth distribution $\Delta = \mathrm{span}\{f_1, \ldots, f_d\}$ is involutive if and only if

$$[f_i, f_j] \in \Delta \ \ \forall \ 1 \leq i, j \leq d. \tag{A.38}$$

References

1. P. Ailer, I. Sánta, G. Szederkényi and K.M. Hangos. Nonlinear model-building of a low-power gas turbine. *Periodica Politechnica, Ser. Transp. Eng.*, 29:117–135, 2002.
2. P. Ailer, G. Szederkényi and K.M. Hangos. Modeling and nonlinear analysis of a low-power gas turbine. Scl-1/2001, Computer and Automation Research Institute, 2001.
3. P. Ailer, G. Szederkényi and K. M. Hangos. Model-based nonlinear control of a low-power gas turbine. In *15th Trieninal World Congress of the International Federation of Automatic Control, Barcelona, Spain*, pages CD–print. 2002.
4. A.A. Alonso, J.R. Banga and I. Sanchez. Passive control design for distributed process systems: Theory and applications. *AIChE Journal*, 46:1593–1606, 2000.
5. A.A. Alonso and B.E. Ydstie. Stabilization of distributed systems using irreversible thermodynamics. *Automatica*, 37:1739–1755, 2001.
6. B.D.O. Anderson and J.B. Moore. *Optimal Control: Linear Quadratic Methods*. Prentice Hall, New York, 1989.
7. P.J. Antsaklis and A.N. Michel. *Linear Systems*. McGraw-Hill, New York, 1997.
8. V.I. Arnold. *Ordinary Differential Equations*. MIT Press, Cambridge, MA, 1973.
9. K.J. Aström and B. Wittenmark. *Computer Controlled Systems*. Prentice Hall, New Jersey, 1990.
10. A. Banos, F. Lamnabhi-Lagarrigue and F.J. France Montoya (eds). *Advances in the Control of Nonlinear Systems (Communication and Control Engineering Series)*. Springer-Verlag, Berlin, Heidelberg, 2001.
11. M. Basseville and I.V. Nikiforov. *Detection of Abrupt Changes. Theory and Practice*. Prentice Hall, London, 1993.
12. L. Bieberbach. *Theorie der Differentialgleichungen*. Springer-Verlag, Berlin, 1930.
13. R.W. Brockett. *Finite Dimensional Linear Systems*. Wiley, New York, 1970.
14. R.W. Brockett. On the algebraic structure of bilinear systems. In R.R. Mohler and A. Ruberti (eds). *Theory and Applications of Variable Structure Systems*, pages 153–168, Academic Press, New York, 1972.
15. R.W. Brockett. Feedback invariants for non-linear systems. In *IFAC Congress*, pages 1115–1120. 1978.
16. C. Bruni, G. Di Pillo and G. Koch. On the mathematical models of bilinear systems. *Richerce di Automatica*, 2:11–26, 1971.
17. C.I. Byrnes and A. Isidori. A frequency domain philosophy for nonlinear systems. *IEEE Conf. Dec. Contr.*, 23:1569–1573, 1984.
18. C.I. Byrnes and A. Isidori. Local stabilization of minimum-phase nonlinear systems. *Syst. Contr. Lett.*, 11:9–17, 1988.

19. C.I. Byrnes, A. Isidori and J.C. Willems. Passivity, feedback equivalence and the global stabilization of minimum-phase nonlinear systems. *IEEE Trans. Aut. Contr.*, AC-36:1228–1240, 1991.

20. C. Chen. *Introduction to Linear Systems Theory*. Holt, Rinehart and Winston, New York, 1970.

21. H. Nijmeijer and A.J. Van der Schaft. *Nonlinear Dynamical Control Systems*. Springer-Verlag, New York, Berlin, 1990.

22. J.C. Doyle. Guaranteed margins for LQG regulators. *IEEE Trans. Aut. Contr.*, AC-23:756–757, 1978.

23. C.A. Farschman, K. Viswanath and B.E. Ydstie. Process systems and inventory control. *AIChE Journal*, 44:1841–1857, 1998.

24. M. Fliess. Matrices de Hankel. *J. Math. Pures Appl.*, 53:197–224, 1974.

25. M. Fliess. Fonctionelles causales non linéaires et indéterminées non commutatives. *Bull. Soc. Math. France*, 109:3–40, 1981.

26. T.R. Fortescue, L.S. Kershenbaum and B.E. Ydstie. Implementation of self-tuning regulators with variable forgetting factors. *Automatica*, 17:831–835, 1981.

27. P. Glansdorff and I. Prigogine. *Thermodynamic theory of structure, stability and fluctuations*. Wiley Interscience, New York, 1971.

28. K.M. Hangos, A.A. Alonso, J. Perkins and B.E. Ydstie. A thermodynamic approach to structural stability of process plants. *AIChE Journal*, 45:802–816, 1999.

29. K.M. Hangos, A.A. Alonso, J.D. Perkins and B.E. Ydstie. A thermodynamical approach to the structural stability of process plants. *AIChE Journal*, 45:802–816, 1999.

30. K.M. Hangos, J. Bokor and G. Szederkényi. Hamiltonian view of process systems. *AIChE Journal*, 47:1819–1831, 2001.

31. K.M. Hangos and I.T. Cameron. The formal description of process modelling assumptions and their implications. In *Proc. PSE-ESCAPE Conference, Comput. Chem. Engng. (Suppl.)*, volume 43, pages 823–828. 1997.

32. K.M. Hangos and I.T. Cameron. *Process Modelling and Model Analysis*. Academic Press, London, 2001.

33. K.M. Hangos, R. Lakner and M. Gerzson. *Intelligent Control Systems: An Introduction with Examples*. Kluwer, New York, 2001.

34. K.M. Hangos and J.D. Perkins. On structural stability of chemical process plants. *AIChE Journal*, 43:1511–1518, 1997.

35. D. Hill and P. Moylan. Connections between finite gain and asymptotic stability. *IEEE Trans. Aut. Contr.*, AC-25:931–936, 1980.

36. A. Isidori, A.J. Krener, C. Gori Giorgi and S. Monaco. Nonlinear decoupling via feedback: a differential geometric approach. *IEEE Trans. Aut. Contr.*, AC-26:331–345, 1981.

37. A. Isidori. *Nonlinear Control Systems*. Springer-Verlag, Berlin, 1995.

38. B. Jakubczyk and W. Respondek. On linearization of control systems. *Bull. Acad. Polonaise Sci. Ser. Sci. Math.*, 28, 1980.

39. T. Kailath. *Linear Systems*. Prentice Hall, New Jersey, 1980.

40. R.E. Kalman. Contributions to the theory of optimal control. *Bol. Soc. Matem. Mex.*, pages 102–119, 1960.

41. R.E. Kalman. The theory of optimal control and the calculus of variations. In R. Bellman (ed), *Mathematical Optimization Techniques*. University of California Press, 1963.

42. H.J. Kreutzer. *Nonequilibrium thermodynamics and its statistical foundations*. Clarendon Press, New York, 1983.

43. C. Kuhlmann, D. Bogle and Z. Chalabi. On the controllability of continuous fermentation processes. *Bioprocess Engineering*, 17:367–374, 1997.

44. C. Kuhlmann, D. Bogle and Z. Chalabi. Robust operation of fed batch fermenters. *Bioprocess Engineering*, 19:53–59, 1998.

45. C. Lesjak and A.J. Krener. The existence and uniqueness of Volterra series for nonlinear systems. *IEEE Trans. Aut. Contr.*, AC-23:1091–1095, 1978.

46. L. Ljung. Recursive least-squares and accelerated convergence in stochastic approximation schemes. *International Journal of Adaptive Control and Signal Processing*, 15:169–178, 2001.

47. L. Ljung and S. Gunnarsson. Adaptation and tracking in system identification – a survey. *Automatica*, 26:7–21, 1990.

48. J.M. Maciejowski. *Multivariable Feedback Design*. Addison-Wesley, Wokingham, UK, 1989.

49. H.A. Nielsen, T.S. Nielsen, A.K. Joensen, H. Madsen and J. Holst. Tracking time-varying coefficient functions. *International Journal of Adaptive Control and Signal Processing*, 14:813–828, 2000.

50. A.V. Oppenheim and R.W. Schafer. *Discrete-Time Signal Processing*. Prentice Hall, New Jersey, 1989.

51. R. Ortega, A.J. Van Der Schaft and B.M. Maschke. Stabilization of port-controlled Hamiltonian systems via energy balancing. *Stability And Stabilization Of Nonlinear Systems*, 246:239–260, 1999.

52. R. Ortega and M.W. Spong. Adaptive motion control of rigid robots: A tutorial. *Automatica*, 25:877–888, 1989.

53. L.S. Pontryagin. *Ordinary Differential Equations*. Addison-Wesley, Reading, MA, 1962.

54. N. Rouche, P. Habets and M. Laloy. *Stability Theory by Liapunov's Direct Method*. Springer-Verlag, New York, 1977.

55. P. Rouchon and Y. Creff. Geometry of the flash dynamics. *Chemical Engineering Science*, 18:3141–3147, 1993.

56. W. Rugh. *Mathematical Description of Linear Systems*. Marcel Dekker, New York, 1975.

57. W. Rugh. *Nonlinear System Theory – The Volterra-Wiener Approach*. The John Hopkins University Press, 1981.

58. E. Feron S. Boyd, L. El Ghaoui and V. Balakrishnan. *Linear Matrix Inequalities in System and Control Theory*. SIAM studies in Applied Mathematics, Philadelphia, 1994.

59. J. La Salle and S. Lefschetz. *Stability by Liapunov's Direct Method*. Academic Press, New York, London, 1961.

60. S. Sastry. *Nonlinear Systems: Analysis, Stability and Control (Interdisciplinary Applied Mathematics/10)*. Springer-Verlag, Berlin, Heidelberg, 1999.

61. S. Shishkin, R. Ortega, D. Hill and A. Loria. On output feedback stabilization of Euler-Lagrange systems with nondissipative forces. *Systems & Control Letters*, 27:315–324, 1996.

62. L. Silverman. Realization of linear dynamical systems. *IEEE Trans. Aut. Contr.*, AC-16:554–568, 1971.

63. H. Siraramirez and I. Angulonunez. Passivity-based control of nonlinear chemical processes. *International Journal Of Control*, 68:971–996, 1997.

64. S. Skogestad and I. Postlethwaite. *Multivariable Feedback Control*. Wiley, Chichester, New York, Toronto, Singapore, 1996.

65. J-J. Slotine. Putting physics in control – the example of robotics. *IEEE Control Systems Magazine*, 8:12–18, 1988.

66. J-J. Slotine and W. Li. *Applied Nonlinear Control*. Prentice Hall, New Jersey, 1990.

67. E.D. Sontag. A 'universal' construction of Artstein's theorem on nonlinear stabilization. *System and Control Letters*, 13:117–123, 1989.
68. E.D. Sontag. *Mathematical Control Theory. Deterministic Finite Dimensional Systems*. Springer-Verlag, New York, 1998.
69. G. Stikkel, J. Bokor and Z. Szabó. Disturbance decoupling problem with stability for LPV systems In *Proc. European Control Conference, Cambridge*, 2003.
70. H. Sussmann. Sufficient condition for local controllability. *Not. Am. Math. Soc.*, 22:A415–A415, 1975.
71. H. Sussmann. Existence and uniqueness of minimal realizations of nonlinear systems. *Math. Syst. Theory*, 10:263–284, 1977.
72. H. Sussmann. Single input observability of continuous time systems. *Math. Syst. Theory*, 10:263–284, 1979.
73. H. Sussmann. Lie brackets and local controllability. *SIAM Journal on Control and Optimization*, 21:686–713, 1983.
74. Z. Szabó, J. Bokor and G. Balas. Detection filter design for LPV systems – a geometric approach In *Proc. 15-th IFAC World Congress on Automatic Control*, 2002.
75. G. Szederkényi, M. Kovács and K.M. Hangos. Reachability of nonlinear fed-batch fermentation processes. *International Journal of Robust and Nonlinear Control*, 12:1109–1124, 2002.
76. G. Szederkényi, N.R. Kristensen, K.M. Hangos and S.B. Jorgensen. Nonlinear analysis and control of a continuous fermentation process. *Computers and Chemical Engineering*, 26:659–670, 2002.
77. F. Szidarovszky and A.T. Bahill. *Linear Systems Theory*. CRC Press, Boca Raton, 1997.
78. F. Szigeti, J. Bokor and A. Edelmayer. Input reconstruction by means of system inversion In *Proc. 15-th IFAC World Congress on Automatic Control*, 2002.
79. A.J. van der Schaft. Representing a nonlinear state space system as a set of higher order differential equations in the inputs and outputs. *Systems Control Lett.*, 12:151–160, 1989.
80. A.J. van der Schaft. *L2-Gain and Passivity Techniques in Nonlinear Control*. Springer-Verlag, Berlin, 2000.
81. M. van der Wal and B. de Jager. A review of methods for input/output selection. *Automatica*, 37:487–510, 2001.
82. E.I. Varga, K.M. Hangos, and F. Szigeti. Controllability and observability of heat exchanger networks in the time varying parameter case. *Control Engineering Practice*, 3:1409–1419, 1995.
83. B.E. Ydstie. Passivity based control via the second law. *Computers and Chemical Engineering*, 26:1037–1048, 2002.
84. B.E. Ydstie and A.A. Alonso. Process systems and passivity via the Claussius Planck inequality. *Systems and Control Letters*, 30:253–264, 1997.

Index